Contents

Part Three: Computing with Quanta

Computing with
Quantum Cats

www.**transworldbooks**.co.uk

Computing with Quantum Cats

From Colossus to Qubits

John Gribbin

BANTAM PRESS

LONDON · TORONTO · SYDNEY · AUCKLAND · JOHANNESBURG

TRANSWORLD PUBLISHERS
61–63 Uxbridge Road, London W5 5SA
A Random House Group Company
www.transworldbooks.co.uk

First published in Great Britain
in 2013 by Bantam Press
an imprint of Transworld Publishers

A CIP catalogue record for this book
is available from the British Library.

ISBNs 9780593071144 (cased)
9780593071151 (tpb)

Addresses for Random House Group Ltd companies outside the UK
can be found at: www.randomhouse.co.uk
The Random House Group Ltd Reg. No. 954009

The Random House Group Limited supports the Forest Stewardship Council®
(FSC®), the leading international forest-certification organisation. Our books carrying
the FSC label are printed on FSC®-certified paper. FSC is the only forest-certification
scheme supported by the leading environmental organisations, including Greenpeace.
Our paper procurement policy can be found at www.randomhouse.co.uk/environment

Typeset in 11.5/16pt Janson by
Falcon Oast Graphic Art Ltd.
Printed and bound in Great Britain by
Clays Ltd, Bungay, Suffolk

2 4 6 8 10 9 7 5 3 1

Acknowledgements

This book grew out of conversations with the quantum computer group at Sussex University, in particular Winfried Hensinger, who opened my eyes to the dramatic progress now being made in the practical application of ideas that seemed esoteric even a few years ago. I already knew something about those esoteric ideas thanks to David Deutsch, of the University of Oxford, and Terry Rudolph, at London's Imperial College. Thanks also to the helpful people at Bletchley Park, Gonville and Caius College, Cambridge, and the David Bohm Archive at Birkbeck College, London; and to John Carl, Frank Carter, Terry Clark, David Darling, Artur Ekert, Lucien Hardy, Mark Hogarth, Betty Houghton, Tero Keski-Valkama, Tony Leggett, Lawrence Lerner, Irfan Siddiqi and Michelle Simmons.

Computing with Quantum Cats

Both experimental and theoretical physicists are currently excited by the prospect of developing computers operating on quantum principles. There is also a lively interest among the military – a source of a great deal of funding – and big business. Quantum computation is one of the hottest scientific topics of the second decade of the twenty-first century, and it all depends on manipulating quantum entities (electrons, photons or single atoms) that are in two states at the same time – exactly like Schrödinger's famous 'dead and alive' cat. Hence my title.

This is a watershed time in computational science, because quantum computers do much more than operate faster than conventional computers – although they certainly do that. For example, they can be used to crack codes that are literally uncrackable by conventional computers, which is a major reason for the interest of the military and big business. This has been known in theory for decades (Richard Feynman was one of the first people to speculate along these lines); but now working quantum computers are actually being used. Admittedly, as yet they involve very large pieces of expensive and temperamental equipment solving very simple problems, such as finding the factors of the number 15. But nobody who has seen the evolution of conventional computers from expensive, temperamental, laboratory-sized machines full of glowing 'valves' to the PC and the iPad can doubt that within a decade the computer world will be turned upside down. More esoterically, such machines will enable physicists to come to grips with the nature of quantum reality, where communication can occur faster than the speed of light, tele-portation is possible, and particles can be in two places at once. The implications are as yet unknowable, but it is fair to say that the quantum computer represents an advance as far beyond the conventional computer as the conventional computer is beyond the abacus.

Conventional computers – often referred to as 'classical' computers – store and manipulate information consisting of binary digits, or bits. These are like ordinary switches that can be in one of two positions, on or off, up or down. The state of a switch is represented by the numbers 0 and 1, and all the activity of a computer involves changing the settings on those switches in the appropriate way. My own computer is, while I am writing these sentences using a word processing program, also playing music, and has an email program running in the background that will alert me if a new message comes in. All of this, and all the other things computers can do, is happening because strings of 0s and 1s are being moved and manipulated inside the 'brain' of the computer.[1]

Eight bits like this make a byte, and because we are

counting in base two rather than base ten the natural steps up the ladder of multiplication do not go 10, 100, 1,000 and so on but 2, 4, 8, 16 and so on. It happens that 2^{10} is 1,024, which is close to 1,000, and we are used to using base 10, so 1,024 bytes is called a kilobyte. Similarly, 1,024 kilobytes make a megabyte, and 1,024 megabytes make a gigabyte. The hard drive of my laptop computer can store 160 gigabytes of information, and the 'brain' – the processor – can manipulate up to 2 gigabytes at a time, all in the form of strings of 0s and 1s (this is now a rather old machine; 'this year's model' can do even better).

But a quantum computer is something else. In the quantum world, entities such as electrons can be in a super-position of states. This means that quantum switches can be in both states, on and off at the same time, like Schrödinger's 'dead and alive' cat. Electrons themselves, for example, have a property called spin, which is not quite the same as what we mean by spin in our everyday world, but can be thought of as meaning that the electron is pointing either up or down. If we say that 'up' corresponds to 0 and 'down' corresponds to 1, we have a binary quantum switch. Under the right circumstances, this switch can exist in a situation where it is pointing both up and down at the same time. Or it can be pointing either up or down, giving three possibilities!

A single quantum switch that is in a superposition of states can 'store' the numbers 0 and 1 simultaneously. By extension from the language of classical computers, such a quantum switch is called a qubit, short for 'quantum bit' and pronounced 'cubit', like the biblical unit of length. The qubits are the 'quantum cats' of my title. The existence of qubits has mind-blowing implications. Two classical bits, for example, can represent any of the four numbers from 0 to 3, because they can exist in any of four combinations: 00, 01, 10 and 11. To represent all four of the numbers (0, 1, 2 and 3) simultaneously, you would need four pairs of numbers – in effect, one byte. But just two qubits can represent all four of these numbers simultaneously. A set of bits (or qubits) operating as a number store in this way is called a register. A register made up of eight qubits (a single qubyte) can represent not four but 2^8 numbers simultaneously. That's 256 numbers stored in a single qubyte. Or, as Oxford physicist David Deutsch would put it, it represents 256 different universes in the Multiverse, sharing the information in some way.

In a functioning quantum computer, any manipulation involving an operation on each of those 256 numbers represented by that qubyte of information is carried out simultaneously in all 256 universes, as if we had 256 separate classical computers each working on one aspect of the problem in our universe, or one computer that had to be run 256 times, once for each value of the number. Looking further into the future, a quantum computer based on a 30-qubit processor would have the equivalent computing power of a conventional machine running at 10 teraflops (trillions of floating-point operations per second) – ten thousand times faster than conventional desktop computers today, which run at speeds measured in gigaflops (billions of floating-point operations per second). These numbers hint at the prodigious power of a quantum computer; but the trick is to find a way of getting useful information out at the end of the calculation – getting the different universes to interfere with one another in the right way to produce an 'answer' that we can understand, without destroying the useful information along the way. This trick is now being achieved by several groups around the world, including a research team at my home base, Sussex University. This book will tell you how, in principle, to build a quantum computer. But to set this in context, I want to go all the way back to the beginning of machine computation as we know it – all the way back to the 1930s, less than a long human lifetime ago, and the work of the man who started the ball rolling.

PART ONE

Card-based census counting machine, 1890.
One of the forerunners of the computer.

CHAPTER ONE

Turing and the Machine

If necessity is the mother of invention, the computer had two mothers – cryptography and the hydrogen bomb. But there was only one father: Alan Mathison Turing.

A child of empire

Turing was conceived in India, where his father, Julius, was a member of the Indian Civil Service helping to administer this jewel in the crown of the British Empire; but he was born, on 23 June 1912, in Maida Vale, London, when his parents were on home leave. He already had a brother, John, born in India on 1 September 1908. When Julius returned to India their mother, Sara,[1] stayed in England with the two boys, but only until September 1913, when she rejoined her husband and left the children in the care of a retired army colonel and his wife, who lived at St Leonards-on-Sea in Sussex. There was a nanny who looked after the two boys and the colonel's four daughters, together with another boy whose parents were overseas, and later three cousins of Alan and John. Their mother returned for the summer of 1915, staying with the boys in rented rooms in St Leonards, and both parents came to England in the spring of 1916 – the first time that Alan really had an opportunity to get to know his father. At the end of this leave, in August, Julius Turing returned to India for his next three years' tour of duty. John had already been sent away to school at Hazelhurst, in Kent; Alan, having been just one of a motley group of children, now became in effect the only child of a single parent, who took him almost everywhere with her, including to the High Anglican church she attended (which he hated) and to art classes (she was an accomplished watercolourist), where he was the darling of the female students.

Alan was remembered as a bright, untidy child with a predilection for inventing his own words, such as 'quockling' to describe the sound of seagulls and 'greasicle' for a guttering candle. It was impossible to pull the wool over his eyes – when his nanny tried to let him win a game they were playing by making poor moves, he saw through the subterfuge and was infuriated; when his mother was reading him a story and left a dull passage out, he yelled: 'You spoil the whole thing.'[2] Nor was he ever in any doubt about the accuracy of his own world-view: he *knew*, for example, that the fruit which tempted Eve in the Garden of Eden was a plum. But he never could tell left from right, and marked his left thumb with a red spot so that he would know which was which.

Having taught himself to read (from a book appropriately called *Reading Without Tears*), Alan first encountered formal education at the age of six, when his mother sent him to a local day school to learn Latin. This failed to stir his interest, but highlighted his great difficulty with the physical process of writing, especially with the ink pens in use at the time. His work was always a mess of scratchy scribbles, crossings-out and blots, reminiscent of nothing so much as the spoof handiwork of Nigel Molesworth in the stories by Geoffrey Willans and Ronald Searle.

Alan's next meeting with his father came in 1919, when Julius's leave included a holiday in Scotland: here the seven-year-old boy impressed his family on a picnic by tracking the flights of wild bees to their intersection to find honey. But in December both parents sailed for India, and Alan returned to the colonel's house in St Leonards while John went back to school in Hazelhurst. The next two years saw a change in Alan. When his mother next returned, in 1921, she found that the vivacious and friendly boy she had left in England had become 'unsociable and dreamy', while his education had been so neglected that at nearly nine he had not learned how to do long division. She took him away for a holiday in Brittany, and then to London, where she taught him long division herself. She later recalled that when he had been shown how to find the square root of a number, he worked out for himself how to find the cube root.

At the beginning of 1922, it was time for Alan to follow his brother John to Hazelhurst, a small school for thirty-six boys aged from nine to thirteen, with just three teachers and a matron who looked after the boys. The brothers were together at Hazelhurst for only one term before John left at Easter for Marlborough College and the public-school education for which 'prep' schools such as Hazelhurst were preparing their boys. The same year, Alan was given a book called *Natural Wonders Every Child Should Know*, by Edwin

Brewster. This first encounter with science made a deep impression on him, especially the way the author likened the workings of the body, even the brain, to a machine. He was less impressed by the sporting activities that young English gentlemen of the time were expected to enjoy (or at least endure), and later claimed that he had learned to run fast (he became a very good long-distance runner as an adult) in order to keep away from the ball during hockey. He was also disturbed by the imprecision of some of his teachers, and wrote to John that one of them 'gave a <u>quite false impression</u> of what is meant by <u>x</u>'. His concern was not for himself, but that the other boys might be misled.

The summer of 1922 brought the return of Alan's father on leave once more, and another happy family holiday in Scotland. But in September his parents left him back at Hazelhurst, departing down the drive of the school with Sara biting her lip as she watched her son running futilely after the taxi, trying to catch up with them. Bored by school, Alan achieved nothing spectacular in the way of marks, but loved inventing things and developed a passion for chemistry – which was purely a hobby: God forbid that a prep school like Hazelhurst should have anything to do with science. Science was almost as conspicuous by its absence at most public schools, so when in the autumn of 1925 Alan surprised everyone by doing well in the Common Entrance examination that was a prerequisite to the transition, his future presented his parents with something of a conundrum. John made an impassioned plea to their parents not to send his unusual younger brother to Marlborough, which 'will crush the life out of him', and Sara Turing worried that her son might 'become a mere intellectual crank' if he failed to adapt to public school life. The puzzle of what to do with him was solved by a friend of hers who was married to a science master at Sherborne, a school in Dorset established back in 1550 and brought into the modern public school system in 1869. The friend assured Sara that this would be a safe haven for her boy, and Alan started there in 1926.

Sherborne

He was due to arrive for the start of the summer term, on 3 May, from Brittany, where his parents were living to avoid paying British income tax. On the ferry to Southampton, Alan learned that there would be no trains, because of the general strike; totally unfazed, and still a month short of his fourteenth birthday, he cycled the 60 miles to Sherborne, staying overnight at Blandford Forum. This initiative was sufficiently unusual to merit a comment in the *Western Gazette* on 14 May. The same initiative and independence were shown a little later when Alan worked out for himself the formula known as 'Gregory's series' for the inverse tangent, unaware that it had been discovered in 1668 by the Scottish mathematician James Gregory (inventor of a kind of telescope that also bears his name), and even earlier by the Indian mathematician Madhava.

Alan soon settled into his old habit of largely ignoring lessons that he found boring, then doing well in examinations, while continuing his private chemistry experiments and amusing himself with advanced mathematics. At Sherborne, grades depended on a combination of continuous assessment and examinations, each marked separately but with a final combined mark. On one occasion, Alan came twenty-second out of twenty-three for his term's work, first in the examinations, and third overall. His headmaster did not approve of such behaviour, and wrote to Alan's father: 'I hope he will not fall between two stools. If he is to stay at a Public School, he must aim at becoming educated. If he is to be solely a Scientific Specialist, he is wasting his time at a Public School.' But Alan escaped expulsion, and was rather grudgingly allowed to take the School Certificate examination, which had to be passed before he could move on to the sixth form at the beginning of 1929. His immediate future after school, however, was decided as much by love as by logic.

As in all public schools, filled with teenage boys with no other outlet for their burgeoning sexuality, there were inevitably liaisons between older and younger pupils, no matter how much such relationships might be officially frowned upon. It was in this environment that Alan realized that he was homosexual, although there is no record of his having any physical relationships with other boys at school. He did, though, develop something more than a crush on a boy a year ahead of him at school, Christopher Morcom.

The attraction was as much mental as physical (indeed, from Morcom's side it was all mental). Morcom was another mathematician, with whom Alan could discuss science, including Einstein's general theory of relativity, astronomy, and quantum mechanics. He was a star pupil who worked hard at school and achieved high grades in examinations, giving Alan, used to taking it easy and relying on brilliance to get him through, something to strive to emulate. The examination they were both working for, the Higher School Certificate (or just 'Higher'), was a prerequisite to moving on to university. In the mathematics paper they sat, Alan scored a respectable 1,033 marks; but Morcom, the elder by a year, scored 1,436.

In 1929, Morcom was to take the examination for a scholarship at Trinity College, Cambridge. He was eighteen, and expected to pass. Alan was desperate not to see his friend go on to Cambridge without him. He decided to take the scholarship examination at the same time, even though he was still only seventeen and Trinity was the top college in Britain (arguably, in the world) for the study of maths and science, with a correspondingly high admission standard. The examinations were held over a week in Cambridge, giving the two Shirburnians a chance to live the life of undergraduates, and to meet new people, including Maurice Pryce, another candidate, whom Alan would meet again when their paths crossed in Princeton a few years later.

The outcome was as Alan had feared. Morcom passed, gaining a scholarship to Trinity that gave him sufficient income to live on as an undergraduate. Alan did not, and faced a separation of at least a year from his first love. But the separation became permanent when Morcom died, of tuberculosis, on 13 February 1930. Alan wrote to his own mother: 'I feel that I shall meet Morcom again somewhere and that there will be some work for us to do together . . . Now that I am left to do it alone I must not let him down.' And in the spirit of doing the work that they might have done together, or that Morcom might have done alone, and 'not letting him down', Alan tried once again for Cambridge in 1930. Once again, he failed to obtain a Trinity scholarship; but this time he was offered a scholarship worth £80 a year at his second choice of college, King's. He started there in 1931, when he was nineteen.

Cambridge . . .

Turing managed the unusual feat of joining in both the sporting life (as a runner and rower) and the academic life in Cambridge, while never quite fitting in anywhere socially. He also enjoyed at least one homosexual relationship, with another maths student, James Atkins. But it is his mathematical work that is important here. Turing's parting gift from Sherborne, in the form of a prize for his work, had been the book *Mathematical Foundations of Quantum Mechanics*, by the Hungarian-born mathematician John von Neumann, who would soon play a personal part in Turing's story.[3] In an echo of his early days at Sherborne, not long after he arrived in Cambridge Turing independently came up with a theorem previously (unbeknown to him) proved by the Polish mathematician Wacław Sierpiński; when Sierpiński's priority was pointed out to him, he was delighted to find that his proof was simpler than that of the Pole. Polish mathematicians would also soon loom large in Turing's life.

In the early 1930s, the structure of the mathematics course in Cambridge was changing. Everybody who entered in 1931 (eighty-five students in all) took two key examinations, Part I at the end of the first year and Part II at the end of the third year. So-called 'Schedule A' students left it at that, which was sufficient to gain them their degrees. But 'Schedule B' students, including Turing, took a further, more advanced, examination, also at the end of their third year. For the intake which followed Turing's year, the extra examination was taken after a further (fourth) year of study, as it has been ever since: it became known as Part III, and is now roughly equivalent to a Master's degree from other universities.

This peculiarity of the Cambridge system partly explains

why Turing never worked for a PhD in Cambridge. Having passed his examinations with flying colours, he was offered a studentship worth £200 which enabled him to stay on at Cambridge for a year to write a dissertation with which he hoped to impress the authorities sufficiently to be awarded a fellowship at King's. In the spring of 1935, still only twenty-two years old, Turing was indeed elected as a Fellow of King's for three years, with the prospect of renewal for at least a further three years, at a stipend of £300 per year; the success was sufficiently remarkable that the boys at Sherborne were given a half-day holiday in his honour. But something much more significant had happened to Turing during his studentship year. He had been introduced to the puzzle of whether it was possible to establish, from some kind of mathematical first principles, whether or not a mathematical statement (such as Fermat's famous Last Theorem) was provable. Apart from the philosophical interest in the problem, if such a technique existed it would save mathematicians from wasting time trying to prove the unprovable.

A very simple example of an unprovable statement is 'this statement is false'. If it is true, then it must be false, and if it is false, it must be true. So it cannot be proved to be either true or false. The mathematical examples are more tricky, for those of us without a Part III in maths, but the principle is still the same. Embarrassingly for mathematicians, it turns out that there are mathematical statements which are true, but cannot be proved to be true, and the question is whether provable statements (equivalent to 'this statement is true') in mathematics can be distinguished from unprovable statements using some set of rules applied in a certain way.

Turing's introduction to these ideas came from a series of

lectures given by Max Newman on 'The Foundations of Mathematics', drawing heavily on the work of the German mathematician David Hilbert. Newman described the application of this kind of set of rules as a 'mechanical process', meaning that it could be carried out by a human being (or a team of such human 'computers') following the rules blindly, without having any deep insight. As the Cambridge mathematician G. H. Hardy had commented, 'it is only the very unsophisticated outsider who imagines that mathematicians make discoveries by turning the handle of some miraculous machine'. But Turing, always idiosyncratic and literal-minded, saw that a 'mechanical process' carried out by a team of people *could* be carried out by a machine, in the everyday sense of the word. And he carried with him the idea, from his childhood reading, of even the human body as a kind of machine. In the early summer of 1935, as he lay in a meadow at Grantchester taking a rest from a long run, something clicked in his mind, and he decided to try to devise a machine that could test the provability of any mathematical statement. By then, he had already met von Neumann, who visited Cambridge in the spring of 1935, and had applied for a visiting fellowship at Princeton, von Neumann's base, for the following year. He would not arrive empty-handed.

Turing came up with the idea of a hypothetical automatic machine that would operate by reading and writing symbols on a long piece of paper – a paper tape. The tape would be divided into squares, and each square would either contain the symbol '1' or be blank, corresponding to the symbol '0'. The way in which the machine was set up would determine its initial 'state'. The tape would start off with a string of 1s and 0s, representing a problem that had to be solved – as Turing was well aware, any information can be represented in such a binary code, if the string of 1s and 0s is long enough.

This may strike you as odd, because the binary 'code' seems so simple. But the printed version of this book, for example, contains a certain amount of information 'stored' in the words of the English language and the letters of the alphabet. It could be translated into binary language simply by setting A = 0, B = 1, C = 10, D = 11 and so on, with extra binary numbers for punctuation marks, and writing out the string of 1s and 0s on a paper tape. Something similar, but not using this particular code, happens when my words are processed by the computer on which I write, at the printer's when the code is turned into printed pages, and, if you are reading an electronic version of the book, inside your reader.

The machine Turing described would, when setting out to solve a specific problem, read the first symbol on a tape, and, in accordance with its state at the time, either erase a 1, print a 1, or do nothing. Then it would move on to the next square, and act in accordance with its new state, which would have been determined by what happened at square 1. It could move forwards and backwards along the tape, but only one square at a time, writing and erasing symbols, until it reached a state corresponding to the end of its task. It would then stop, and the string of 1s and 0s left on the tape would represent the solution to the problem the machine had been working on. And it would have been done by a purely 'mechanical' process, owing nothing to inspiration or human intuition.

In terms of the original problem he had set out to solve, Hilbert's provability question, Turing's hypothetical machine was a great success. Simply by considering the way in which such a machine would work, he was able to show, using a detailed argument which we do not need to go into here, that there are uncomputable problems, and that there is no way to distinguish provable statements in mathematics from unprovable statements in mathematics using some set of rules applied in a certain way. This was impressive enough. But what is even more impressive, and the main reason why Turing's paper 'On Computable Numbers' is held in such awe today, is that he realized that his 'automatic machine' could be a universal computer. The way the machine works on a particular problem depends on its initial state. It is a limited machine that can only solve a single problem. But as Turing appreciated, the initial state can be set up by the machine reading a string of 1s and 0s from a tape – what we now call a computer program. The same piece of machinery (what we now call hardware) could be made to do any possible task in accordance with appropriate sets of instructions (what we now call software). One such machine can simulate the work of any such machine. And such machines are now known as Turing machines. In his own words, 'it is possible to invent a single machine which can be used to compute any computable sequence'.

The relevance of this idea to the logical puzzle that triggered Turing's investigation is that although he proved that it is possible to construct a machine to solve any solvable problem, it is not possible to construct a machine which can predict how many steps it will take to solve a particular problem. This is what establishes that although you can build an automaton to do anything that can be done, you cannot build a machine which tells you what can and can't be done. Logicians appreciate the full importance of this proof; but that is less important to us here than the fact that Turing machines exist.

A Turing machine simulates the activity of any specialist computer, using different sets of software. This is exactly what my iPhone, for example, does. It can be a phone, a TV or a navigation aid; it can play chess, solve certain kinds of mathematical problems, and do many other things. It can even do things its designers never thought of, as when an outside programmer devises a new app. Most people in the developed world now own, or have access to, a Turing machine, less than eighty years after the publication of 'On Computable Numbers'.

The paper was completed in the spring of 1936, just after the German army re-occupied the Rhineland, and it was published just under a year later, in the *Proceedings of the London Mathematical Society*. In the interim, an inconvenient blip occurred. Just a month after he had read an early draft of Turing's paper, Max Newman received a copy of a paper by Alonzo Church, a mathematician based at Princeton, in which he reached the same conclusion about Hilbert's question, using a technique he called lambda calculus. In a sense, Turing had been pre-empted, and although his version was still worth publishing, he had to add an appendix establishing that his work and Church's work were equivalent. Nobody realized, at the time, that the really important discovery described in that paper was the principle of the universal Turing machine.

. . . and Princeton

Encouraged by Newman and the possibility of working with Church, Turing was determined to make his visit to Princeton. He had applied for one of the Proctor Fellowships offered by Princeton; there were three of these each year, one for a

Cambridge scholar, one for Oxford and one for the Collège de France. Turing's application was unsuccessful – the Cambridge award that year went to the astronomer and mathematician Raymond Lyttleton – but he decided he could scrape by on his King's fellowship, which continued even when he was away, and went anyway, sailing from Southampton on the liner *Berengaria* on 23 September 1936.

'Working with Church' didn't quite live up to expectations, although the pair got on well enough, by the standards of Church's interactions with other people. Church was one of those people, not uncommon in the mathematical sciences, with a tendency towards autism – the physicist Paul Dirac, described by his biographer Graham Farmelo as 'the strangest man', is a prime example, although Church seems to have been nearly as strange. A colleague described him as speaking 'slowly in complete paragraphs which seemed to have been read out of a book, evenly and slowly enunciated, as by a talking machine'.[4] His lectures always began with a ritual cleaning of the blackboard with soap and water, followed by ten minutes waiting for it to dry; but since the lectures simply consisted of reading out the same typewritten texts every year, there was plenty of time to spare. In 1936, Church was thirty-three and Turing twenty-four. In their own ways, both were a little peculiar and used to working alone. Turing was desperately shy, and had a fierce stammer (which, curiously, never troubled him when he was reading from a prepared script for radio broadcasts). It is scarcely surprising that there was no significant collaboration between them, although they did work together on a paper spelling out the equivalence of their two approaches to the Hilbert question, and Church acted as the formal supervisor for a thesis Turing wrote to obtain a Princeton PhD, although by then he hardly needed the degree. More significantly, Turing established contact with von Neumann, of whom more shortly, who certainly appreciated the implications of 'On Computable Numbers'. According to the computer scientist Julian Bigelow, who was in Princeton at the time, the fact that it was possible in principle to build a universal machine that could imitate all other machines 'was understood by von Neumann'.[5] When Turing decided to apply again for a Proctor Fellowship to enable him to spend a second year at Princeton, it was von Neumann who wrote a letter of recommendation on his behalf. Turing, he said, 'has done good work in branches of mathematics in which I am interested', and was 'a most deserving candidate'. With such support, this time the application was successful. Turing spent the summer of 1937 in England, where he got to know the philosopher Ludwig Wittgenstein, before returning to the United States 'a rich man', as he told his mother, with $2,000 to live on.

Unlike Church, Turing had practical skills. He had become interested in cryptography, specifically the kind of codes (strictly speaking, ciphers, but I shall use the terms interchangeably) that involve translating messages into numbers and then manipulating the numbers to produce a coded message. In a simple example, you can represent the letter A by 1, B by 2 and so on. Once your message is written out as a string of numbers, you then multiply that number by a large prime number to produce a new number, which can be transmitted openly. The person receiving the message can decipher it by dividing by the original prime number, but nobody who does not know this 'key' can read the message. Back in Princeton, Turing decided to build a machine to carry out this kind of multiplication, on a very small scale. He was motivated by increasing concern about the prospect of war in Europe, and told Malcolm MacPhail, a Canadian physicist who lent Turing a key to the workshop at Princeton, that the ultimate object was to produce a cipher that would require '100 Germans working eight hours a day on desk calculators 100 years to discover the secret factor'.[6]

The machine used electromechanical switches – relays like those used in telephone exchanges at the time, controlled electrically. Such switches can only be either on or off, representing the 1s and 0s of a binary number. And binary multiplication is particularly simple, since it goes '0 × 0 = 0, 1 × 0 = 0, 1 × 1 = 1' – and that's it. Turing's electronic multiplier was never finished; it was only a part-time project, with limited resources. But he did build several components of the machine which worked satisfactorily.

In March 1938, the same month that Germany swallowed up Austria in an *Anschluss*, Turing was re-elected as a fellow of King's, although the news did not reach him immediately. Before it did, his father wrote urging him to find a permanent post in America, safely out of the way of any conflict, and Turing was actually offered a job as von Neumann's assistant at the Princeton Institute for Advanced Study (IAS), at a salary of $1,500 a year. But he was eager to return home, and never seriously considered the offer. He arrived back at Southampton on 23 July 1938, armed with a fresh (but pointless) Princeton PhD and the pieces of his electronic multiplier wrapped in brown paper. Almost immediately, he was recruited to take a summer course at the Government Code and Cipher School (GC&CS), then based in London, as one of several people identified through the old boy network as likely to be useful in that branch of wartime intelligence.

Not long after this, Turing had another experience that made a deep impression on him. In Cambridge that October he saw the Disney film *Snow White and the Seven Dwarfs*, and became fascinated by the scene in which the White Witch prepares the apple for Snow White. He used to go around reciting her incantation:

> Dip the apple in the brew,
> Let the Sleeping Death seep through.

Bletchley and the Bombe

GC&CS had become aware of the need to recruit mathematicians in case war should break out largely because of the adoption by the German military of a cipher machine called Enigma. The version of Enigma in use at that time contained three moveable discs (called rotors) in a line and a fourth, fixed, 'reflector', each with twenty-six electrical contacts, one for each letter of the alphabet. An electronic signal produced by pressing an individual letter on a keyboard (A, perhaps) passed into the first rotor and out through a *different* contact (maybe corresponding to L) into the adjacent contact in rotor two, and so on for rotors two and three, following a path determined by the alignment of the rotors *and* their internal wiring; then it was scrambled once more before being reflected back from the fourth rotor. After another three stages of scrambling, going back through the three rotors in reverse order, it then lit up a light bulb corresponding to another letter of the alphabet (crucially, *never* the same letter as the original keystroke). The operator then entered this as the first letter of the coded message, and proceeded to the next – but as that next key was pressed, the rotors moved on one click, so that a different electrical path would now be followed, and if the operator pressed A twice in a row, for example, two different letters would appear in the coded message.

The great practical advantage of these machines, thanks to the reflection process, was that, assuming the rotors of two machines were set up in the same way to start with, in order to decipher the message coded on one machine and broadcast by, say, Morse code, an operator far away using the other machine had only to type in the coded message, letter by letter, to retrieve the original plain text. The military version of Enigma also included a development called a plugboard, which involved literally plugging wires into pairs of holes in a board to connect pairs of letters at the first stage of the coding process. If, for example, J and G were joined in this way, and also Q and B, pressing key J would send the signal through contact G on the first rotor, not through J, and back perhaps as the coded letter Q which would follow the plugboard to emerge as B. Fortunately for the codebreakers, only six or seven pairs of letters were usually connected in this way.

There are $26 \times 26 \times 26 = 17{,}576$ different ways to connect the three rotors in such a machine (the reflector does not rotate), and the code for a particular message could be cracked (if you knew the wiring of the rotors themselves) simply (if tediously) by trying out all 17,576 combinations to find the one that worked. The three rotors could be arranged in six different orders (1,2,3; 1,3,2; 2,1,3; 2,3,1; 3,1,2; 3,2,1), but even $6 \times 17{,}576$ is not a number to daunt cryptanalysts. A plugboard with seven possible pairs out of 26 letters, though, introduced 1,305,093,289,500 ways of setting the machine for *each* of the 6 × 17,576 rotor settings. This was such a high number that the Germans were convinced that the Enigma code was uncrackable, and initially the British, sharing that view, devoted little attention to it. But, through a combination of luck and brilliance, the Poles, deeply concerned about the threat to their country from the Nazis, found a way into Enigma.

The luck came in 1932, when French spies got hold of a set of instructions which could be interpreted to reveal the wiring of the Enigma rotors. The French shared this information with their allies Poland and Britain, but only the Poles had the initiative to set a team of mathematicians the task of interpreting the information in this way. The skill came first in unravelling the wiring, and then in breaking the pattern used by the Germans to set up the Enigma machines each day. These basic settings were laid down in a set of instructions issued to all operators, and known as a 'ground setting'. This would involve arranging the order of the rotors, then adjusting the three rotors, rather like setting a combination lock, so that a particular set of three letters, such as BKW, was at the top. Using this setting, the operator would choose his own setting for the rotors, perhaps XAF, and encipher this *twice*, producing a six-letter string such as AZQGBP, and send this on the ground setting before turning the rotors to the chosen setting and enciphering his message.

The snag was that each operator was using the same triplet, coded twice, based on the same rotor setting, at the start of every day – so that hundreds of short messages with identical content were going out coded the same way. The system was later improved so that although all operators used the same rotor setup each day, each operator could choose the initial setting and transmit it uncoded, in plain language, before going through the rest of the setup procedure. Even with this refinement, the repetition involved, and details such as the fact that no letter could be coded as itself, produced patterns when many messages were analysed, and this enabled the Poles to draw up statistical tables from which they could eliminate most possible settings of the Enigma machines each day and end up with the right setting. This was still a time-consuming process to carry out by hand. But the great breakthrough came when the Poles devised an electro-mechanical machine, using commercially available relays like those in Turing's prototype multiplier, to work through all the possibilities. The relays clicking inside the machine when it was working made a sound like the ticking of the clockwork mechanism of a time bomb, so the machines became known as Bombas to the Poles, and more advanced machines, essentially developed by Turing and superficially similar to the Polish devices, were later dubbed Bombes by the British.

Until the end of 1938, with the aid of their Bombas the Poles were cracking the German Enigma codes not because Enigma was inherently unsafe, but because the Germans, complacently sure it was uncrackable, were careless in its use. This would be a recurring theme: what should have been an uncrackable system (and was, when used properly – by the German navy, in particular) was compromised by foolishness at high level, as in the use of repeated triplets just described, and personal stupidity at lower levels, as when operators began or signed off their messages 'Heil Hitler' or used a girl's name for their initial rotor setting.

It was further made vulnerable by the carelessness of its operators and the bureaucratic nature of their system, so that many messages might begin with the German words for 'Daily Report', for example. One way into a message would be to take a 'crib' for a common word, such as *Flugzeug* (aeroplane) and find a match by 'dragging' a string of letters corresponding to the word through a message;[7] and there were other statistical techniques. This process was greatly simplified by the German habit of using standard phrasing: for example, starting weather reports with the words *Wetter für die Nacht* ('weather for the night') and instructions sent out to Luftwaffe squadrons with the phrase 'special instructions for' followed by the squadron number. But all that only just made the Enigma codes crackable, and the cryptanalysts' work often suffered setbacks – as at the end of 1938, when the Germans introduced two new rotors for each machine, making sets of five, from which three were chosen each day. Instead of there being six ways to order the rotors actually used, there were now sixty possibilities, and even with the aid of their Bombas the Poles could no longer cope. Worse, early in 1939 the Germans increased the number of pairings on the plugboard from six to ten. This was the situation in the summer of 1939 when, with war looming, the British and French sent teams to Warsaw to discuss the situation, and were astonished when the Poles revealed what they had managed to achieve.

Not the least of those achievements was that the Poles managed to keep the secret of the Bombas and the cracking of Enigma from their German occupiers after their country was invaded in September 1939. By then, GC&CS had been moved to a country house in Buckinghamshire, Bletchley Park, where Turing was among the prospective codebreakers ordered to report on 4 September.[8] There, he was instrumental in the design and manufacture of the British Bombes, much more sophisticated machines capable of dealing even with the five-rotor system and the ten pairs of plugboard settings, provided the German operators of Enigma were careless enough to provide scraps of 'cribs' such as weather reports headed *Wetter*, girls' names used for ring settings, and at least one occasion when an operator had to repeat a long message and sent it twice using the same rotor settings. Without such cribs, the task of cracking Enigma would have been impossible; even with the cribs, it was horrendously difficult. The British Bombes – which, in spite of their name, were different from and far superior to the Polish Bombas which were their inspiration – each stood nearly 7 feet high, was a full 7 feet wide, and weighed a ton. Each simulated the effect of thirty Enigma machines (later versions were essentially thirty-six Enigmas wired together) working at once through all the possibilities for a particular message. And they only worked at all because of Alan Turing. There is no need to go into all the details here, but as Simon Singh has summed it up, 'only Turing, with his unique background in mathematical machines, could ever have come up with [the British Bombe]'.[9] Turing worked out the logic of a system whereby the setup of the Enigma machines on a particular day could be worked out using cribs; the genius of his system was that instead of working through each possible setup until the codebreakers hit on the right one (by which time the setup might have been changed), he showed that it was possible to find all the wrong answers at once, leaving the correct setup clear by default. His logic was translated into the mechanics of the British Bombes by the British Tabulating Machine Company, based at Letchworth, in

Hertfordshire; a key refinement of Turing's technique, speeding up the codebreaking significantly, was made by his colleague Gordon Welchman.

It has been estimated by sober military historians that through the success of this project Turing was personally responsible for shortening the war by two years. He may even have been responsible for keeping Britain in the war at all; in the summer of 1941, after a desperate period when shipping was being sunk at such a rate that the country was on the brink of starvation, thanks solely to the codebreakers at Bletchley and their Bombes there was a period of twenty-three days without a single sinking, because convoys were being routed away from known U-boat positions. But the entire Bletchley Park effort was kept under wraps, and details only emerged decades later. Turing never spoke about it – not so much because he was bound by the Official Secrets Act, but because he had promised not to speak about it, and to Turing promises were never made lightly and always to be kept.

Many of the stories about Turing's eccentricity date from his time at Bletchley Park. Some of these seem not unreasonable, such as his habit of cycling to work wearing a gas mask during the hay-fever season. Others, such as the time he was hauled over the coals for failing to sign his identity card and replied that he had been told not to write anything on the document, may have owed as much to bloody-mindedness as to an autistic literalism. In order to learn how to shoot, Turing joined the Home Guard, and when confronted with a form which asked, among other things, 'Do you understand that by enrolling in the Home Guard you place yourself liable to military law?' wrote 'No' in the appropriate space and went on to the next question. Having learned to shoot, Turing stopped attending the Home Guard, and was summoned before a Colonel Fillingham who warned him that he was subject to military law and could not pick and choose when to attend parade. Turing calmly explained the situation, the form was dug out from the files, and the authorities were forced to admit that he had never actually been a member of the Home Guard at all. Turing had got what he wanted, by being scrupulously honest and upfront. If others chose to make mistakes, then, as with the careless German Enigma operators, that was their lookout.

Turing's mother recalls an incident which, although she could not have known it when she wrote her book, showed how his approach to problem-solving in everyday life resembled his approach to codebreaking. While at Bletchley, Turing had a bicycle with a fault which made the chain come off after a certain number of revolutions of the pedals. In order to avoid having to stop and re-fix the chain, he first hit on the approach of counting the number of turns of the pedals so that he could make a suitable jiggle at the right time to stop it jumping off its sprocket. Then, he fitted a counter on the bicycle to keep track of the rotation of the wheels for him. Finally, he discovered that there was a mathematical relationship between the number of rotations of the pedals, the number of turns of the wheel and the number of spokes in the wheel. The statistics revealed that the problem was caused by a bent spoke coming into contact with a slightly damaged link in the chain at regular intervals. When the spoke was straightened, the problem was solved. As Sara Turing says, 'a bicycle mechanic could have fixed it in five minutes'. But Turing's approach was entirely logical – especially to a codebreaker involved in analysing the wheel patterns of Enigma machines.

One of the more curious aspects of Turing's life while working at Bletchley Park was that in 1941 he got engaged, to Joan Clarke, a mathematician and colleague. It was only after Joan had accepted his proposal of marriage that Alan told her about his 'homosexual tendencies', but she seemed entirely unworried, and the relationship continued as a close friendship until Alan decided that he could not go through with the charade, and broke the engagement off. At the time, nobody except Joan knew the real reason why.

Although he was instrumental in the success of the Bombe, Turing did not play a central role in the development of its successor, the first electronic computer, Colossus. His later contribution to the war effort took him in different directions – first, in November 1942, to the United States, sailing on the *Queen Elizabeth* to bring the Americans up to speed on the codebreaking work being carried out in the UK. He met up with his cryptographic counterparts in the US Navy's 'Communications Supplementary Services (Washington)' branch, or CSAW, then moved on to the Bell Laboratories, at the time part of the American Telephone and Telegraph Company (AT&T), where he became engrossed in the problem of 'scrambling' speech, so that voice conversations could be transmitted in a form that could not be deciphered without the right equipment. There, Turing met Claude Shannon. They were working on separate secret projects, and could not discuss their war work with one another; but they discovered a shared interest in the possibility of thinking machines, and encouraged each other to speculate about the possibilities. Lunching with Shannon in the executive dining room one day, Turing brought the hubbub of conversation in the room to a halt by declaring loudly to his friend: 'I'm not interested in developing a *powerful* brain. All I'm after is just a *mediocre* brain, something like the President of the American Telephone and Telegraph Company.' And he went on to consider the possibility of a computer that would follow the stock market and give advice on when to buy or sell.

There may have been more to this than unthinking honesty. John Turing says that although his brother could not stand social chat, 'what he really liked was a thoroughly disputatious exchange of views', and if 'you ventured on some self-evident proposition, as, for example, that the earth was round, Alan would produce a great deal of incontrovertible evidence to prove that it was almost certainly flat'.

It is likely that Turing visited Princeton during his wartime travels in the United States; his mother recalled his mentioning such a trip, but there is no official record of such a visit. In the spring of 1943 he returned to Britain. While he had been away, the key turning point of the war in Europe occurred with the surrender of the German forces at Stalingrad on 2 February 1943. But this did nothing to reduce the risks of travelling by ship across the North Atlantic, where the U-boats were still very active. Turing sailed on the *Empress of Scotland* on 23 March, nine days after the *Empress of Canada* had become one of their many victims; he might easily have been on the earlier ship.

Back in Britain, Turing's work concentrated on the speech encipherment system, codenamed Delilah, which would eventually work, but too late to play a part in the war effort.[10] This project was based not at Bletchley, but at a nearby secret centre, Hanslope Park. So Turing was also physically distanced (if only by about 10 miles) from the new hardware developments at Bletchley. But his fingerprints were all over the techniques used by the Bletchley team, and he would re-engage with the fruits of their labours after the war.

The flowering of Colossus

In the summer of 1941, the British intercepted a new kind of radio traffic, codenamed Tunny, operating initially between Berlin and Greece. This was an experimental link which operated until October 1942, when it was modified and began to appear on other routes, including those between Berlin and the German forces in Russia, from Berlin to Rome and North Africa, and to Paris. It emerged that this was being used for high-grade information, including direct orders from Hitler; a potential gold mine for the British and their allies. But Tunny was very different from Enigma, and even harder to crack.

The first difference was that Tunny used teleprinter language, rather than Morse code. This was not a problem in itself, but needs some explanation. Instead of the strings of dots and dashes produced by Morse, teleprinters represent letters of the alphabet and a few punctuation marks in terms of groups of five 'on or off' symbols. These symbols could be represented by holes punched across the width of a paper tape, one-and-a-half inches wide, where a hole meant 'on' and no hole meant 'off'. These symbols were usually represented by o and x, so that a single letter in teleprinter language might read xxoxo, and so on. This is exactly equivalent to a five-bit binary language, in which the same letter would be represented by 00101. A series of these 'letters' was punched automatically along the tape as an operator typed on a teleprinter machine, which was rather like a typewriter. The tape could then be fed into a transmitter and run at high speed, broadcasting the message in a concentrated burst of radio transmission. At the other end, the incoming message was read automatically by the receiving apparatus and used to punch out the holes in another strip of paper tape, which could be fed into a teleprinter machine to print out the message.[11] That message would, of course, be quite transparent, since the teleprinter language was no secret.

The coding for Tunny involved a machine superficially like an Enigma machine, but much more complex. For a start, a Tunny machine contained twelve wheels, each of which could rotate to a different number of positions: 43 for the first wheel, then 47, 51, 53, 59, 37, 61, 41, 31, 29, 26 and 23. This odd-looking pattern was carefully chosen so that the numbers were 'relatively prime', which means that no more than one of them can be divided by any number except 1. So 26, for example, can be divided by 13 and 2, but none of the other numbers can be divided by either 13 or 2. This was a way to avoid certain statistical patterns emerging as the wheels rotated at different rates. When the operator pressed the lever for a letter, all the wheels in the machine worked together to produce another letter, called the key, which was then added to the original letter to produce the encrypted letter of the message. The wheels then moved on in a certain way before encrypting the next letter.

This process of adding letters to one another is easy in binary language, where o + o = o, x + x = o, x + o = x, and o + x = x. So adding xxoxo and oxxox would give you xoxxx. And in a neat twist, adding the same key again restores the original message! So provided the Tunny machine at the other end had the same wheel settings, it would subtract out the key (by adding it again) to leave the message. In a useful but not essential refinement, Tunny did all this automatically, letter by letter as the operator typed or as a paper tape ran through the machine. But the system did have a weakness: in the first version of Tunny the operator had to transmit a string of twelve letters to tell his opposite number the initial wheel settings for the message that followed.

Since the British had never seen a Tunny machine and did not know what went on inside it, this should not have mattered. But in August 1941 they intercepted two Tunny transmissions each preceded by the same code, HQIBPEX-EZMUG, and followed by a message just under four thousand characters long. In an astonishing lapse, an operator had sent the same message twice, using the same wheel settings, which meant with the same key. Just as adding the key twice leaves the original message intact, so adding the encrypted message to itself leaves the key intact. By adding the two messages together and doing some further manipulation they were left with a key 3,976 characters long, which contained information about the encrypting process going on inside the machine. In one of the most impressive achievements of the entire Bletchley Park effort, Bill Tutte, a mathematician from Cambridge, was able, with assistance from his colleagues, to work out the entire structure and operation of the Tunny machine by analysing the statistical patterns in this key. When the Tunny system changed in October 1942 so that the wheel settings were no longer being broadcast clear but were based on predetermined arrange-ments unknown to the codebreakers, at least the Bletchley people knew what they were confronted with.

Tunny should still have been unbreakable, but like

Enigma it was made vulnerable by the carelessness of its operators and the bureaucratic nature of their system. The greatest gifts to the codebreakers were messages repeated without the wheel settings being changed; these were known as 'depths'. Making full use of such carelessness, the code-breaking depended on using techniques developed by Tutte to obtain some of the key wheel settings. This involved very straightforward but tedious calculations.

It was Turing who developed the methods by which the messages could actually be read, once the workings of the machine were understood. These produced good results until the Germans tightened their security, but became significantly harder to apply as time passed and mistakes such as depths became rarer. The technique still worked, but the problem was that these methods were intensely labour intensive and slow. To paraphrase Turing, it was getting to the point where it would require '100 Britons working eight hours a day on desk calculators 100 years to discover the secret factor'. By the end of 1942, it was appreciated that the only way to tackle the problem was with a machine. This approach was suggested by Max Newman, Turing's former teacher in Cambridge, who had been recruited to Bletchley Park a few months earlier, and was now put in charge of the project.

The prototype machine that began operating in June 1943 came to be known as Heath Robinson, because of its bizarrely complex appearance – after a cartoonist of the time who specialized in intricate drawings of complex fantasy machines to do simple things like boiling an egg. Bletchley's Heath Robinson could read two long loops of paper tape at once, using photoelectric detectors and light passing through the holes in the tape. One tape contained an encoded message to be broken, the other a 'code' containing all the possible settings of one group of wheels in the Tunny machine known as chi-wheels. The machine compared the possible settings of chi with the message, one by one, using electronic counters to record the number of hits, until it found a match. Once the chi-wheels were broken, the cryptographers could tackle the message by hand, using cribs, dragging and so on.

Heath Robinson worked after a fashion, but it was slow (limited by the speed with which paper tape could be read), prone to breakdowns when the tape stretched (making it impossible to keep the two tapes in synchrony) or broke, and not entirely reliable (sometimes it would give different results if set the same problem twice). But it proved that the machine approach to breaking Tunny could work. What was needed was a better machine, and by great good fortune the man Bletchley Park asked to build a better machine was exactly the right man for the job.

Among those who had worked on the construction of Heath Robinson were engineers at the Post Office research station in Dollis Hill in north London, who knew all about relays from their work on automatic telephone exchanges. The top engineer at Dollis Hill was Thomas Flowers (known as 'Tommy' at the time, although he preferred 'Tom' in later life). Born in London's East End in 1905, Flowers was the son of a bricklayer, and a genuine Cockney. He had won a scholarship to a technical college, and then joined the Post Office as a trainee telephone engineer, continuing his studies at evening classes and earning a post at Dollis Hill in 1930. There, he pioneered the use of electronic valves for switching in the 1930s, flying in the face of the received wisdom that such valves were unreliable and prone to break down. He had found that the problems arose when valves were repeatedly turned on and off, but that if they were left on all the time, glowing like little incandescent light bulbs, they would run reliably for a very long time without burning out. As early as 1934, he had worked on an experimental telephone switching system using four thousand valves, and a design based on his work had just started to come into operation at the beginning of the war. Flowers himself, though, very nearly spent the war interned in Germany. He was working in Berlin in the late summer of 1939, but fortunately was warned by the British Embassy to go home, and crossed the border into Holland a few hours before the frontier was closed.

Flowers was asked to help with Heath Robinson because Turing had discussed with him the possibility of building an electronic version of the Bombe; although this never happened, Turing was impressed by the engineer and recommended him to Newman as the right man to fix the problems with Heath Robinson. But when he was asked for his advice on how to make the relays in Heath Robinson more reliable, Flowers' reply was that the best thing to do would be to forget about mechanical relays altogether, and use valves instead.

The idea of a reliable machine using a couple of thousand electronic valves was regarded as a fantasy by Newman and his colleagues, who doubted that even if it could be built it would be working in time to contribute to the war effort (this was in February 1943). Flowers was told that he was welcome to try once he was back at Dollis Hill, and in the meantime, rather than officially encouraging the project, Newman ordered another dozen Heath Robinson type machines. But the Director of the Dollis Hill research station, W. G. Radley, saw the potential of the idea (and had first-hand knowledge of

Flowers' success with valve-based machines) and gave his full support to the enterprise (moral support, that is: funds were limited, and Flowers had to pay for some of the equipment himself). The result was a prototype machine, dubbed Colossus, which used 1,600 valves and required only one paper tape, carrying the message to be broken, as the 'chi-stream' to be tested – all the possible settings of the chi-wheels – was generated electronically. After a heroic round-the-clock effort by Flowers and his colleagues, Colossus was tested at Dollis Hill in December 1943, then disassembled and taken on lorries to Bletchley Park, where it arrived on 18 January 1944. Re-assembled, it filled a whole room. The Bletchley Park codebreakers, including Newman, were astonished: 'I don't think they [had] really understood what I was saying in detail – I am sure they didn't – because when the first machine was constructed and working, they obviously were taken aback. They just couldn't believe it! . . . I don't think they understood very clearly what I was proposing until they actually had the machine.'[12]

The re-assembled Colossus broke its first message on 5 February 1944. It was ten times faster than Heath Robinson, and, equally important, more reliable. Orders for more Robinsons were cancelled, and Flowers was asked how quickly Dollis Hill could produce more 'Colossi'.

One of the Wrens[13] who worked on Colossus, Betty Houghton (née Bowden), now lives in a neighbouring village to us. She was fourteen when the war broke out, and three years later joined the WRNS. She was told that there were two kinds of job available – cook/steward or 'P5'. Having no wish to be a cook/steward, she asked what P5 was. 'That's secret,' she was told, and promptly volunteered. She ended up as a Watch Leader in Hut 8 at Bletchley Park, working on Tunny, and recalls Turing as 'a very nice man, very quiet; a bit daft, like most of them'.

Colossus was the first electronic computer. It was also programmable, in a limited sense, because Flowers had deliberately designed it so that it could be adapted to new requirements by switches, and by plugging cables linking the logic units in different arrangements. The crucial difference from a modern computer, though, is that it did not store programs in its memory, the way Turing had envisaged; the programming had to be done literally 'by hand' at the switches and plugboards. Even so, this adaptability proved an enormous asset, and Colossi could be adapted to use new codebreaking methods as they were invented, carrying out tasks that its designer could not have imagined.

Flowers was asked to have an improved Colossus up and running at Bletchley by 1 June 1944. He was not told why, but the urgency was stressed. The tight deadline was met by having the machine, containing 2,400 valves and running 125 times faster than electromechanical machines, assembled and tested on site. It began operating on 1 June, as requested; although Flowers did not know it at the time, this was intended to be D-Day, the date of the invasion of German-occupied France. Bad weather delayed the invasion, and as it continued there were serious doubts about whether the Allies would be able to ship enough men and materiel across the rough English Channel to support the invasion against a counter-attack. But on 5 June Colossus II was instrumental in breaking a message which revealed that Hitler had completely fallen for the Allied deception plan (Operation Fortitude), which led him to believe that the invasion would strike at the

Pas de Calais, with a diversionary raid in Normandy. In the intercepted Tunny signal, he ordered Rommel to hold his forces in the Pas de Calais area to repel the 'real' invasion, due five days after the expected Normandy landing. It was this piece of information, combined with a forecast of slightly improving weather, that clinched Eisenhower's decision to go ahead on 6 June, knowing that even in bad weather five days would give his forces time to build up the beachhead.

By the end of the war in 1945 eight more Colossi had been installed at Bletchley Park, and Eisenhower himself later said that without the work of the codebreakers the war would have lasted at least two years longer than it did. The two men who did more than anyone else to make all this possible were Turing and Flowers. They should each have been knighted at the end of hostilities, and given every support to develop their ideas further. But that isn't the way it happened.

Anticlimax: after Bletchley

Harry Fenson, a member of Flowers' team, has said that he was well aware at the time that Colossus was 'a data processor rather than a mere calculator, and rich in logical facilities'. It had the potential to manipulate many kinds of data, 'such as text, pictures, movement, or anything which could be given a value'. It contained 'all the elements to make a general-purpose device' – a Turing machine.[14]

At the end of the war, Bletchley Park could (and I would say, should) have become a scientific research centre, equipped with ten Colossi and a world lead in computing. Instead, on the direct orders of Winston Churchill (who did many questionable things to set alongside his greater moments), all but two of the machines were physically broken up and most of their components smashed. This was part of a successful attempt to hide the success of the codebreaking work which had substantially shortened the war, so that the British could carry on reading the coded traffic of other nations without being suspected. The 'other nations' included the Soviet Union, which used captured German Tunny machines long after the war. In April 1946, the codebreaking headquarters moved to Eastcote, a London suburb, and changed its name to the Government Communications Headquarters (GCHQ); GCHQ moved on to Cheltenham, its present home, in 1952. In both these moves, it took with it the two remaining Colossi ('Colossus Blue' and 'Colossus Red'); the work they did is still classified. One was dismantled in 1959, the other in 1960. But all is not quite lost; a replica Colossus has been built at Bletchley Park, which is now a museum, and can be seen there in all its glory.

As machinery was physically destroyed, so papers were burned and the codebreakers were all sworn to secrecy – and they all kept their secrets, in many cases taking them to the grave. The attitude that wartime secrets should not be enquired into was shared by people outside Bletchley Park. When I asked Betty Houghton what she had said to her parents when they enquired about her war work, she replied, 'They never asked.' The story of Enigma did not emerge properly until the 1970s, and that of Colossus became known in detail only after a crucial document, called *General Report on Tunny* and written in 1945, was released in 1996 under the American Freedom of Information Act. Deliciously, it is now available online to anyone with a Turing machine.[15]

Tom Flowers, the man who designed and built the first electronic computer, never imagined that the secrecy would last so long. Although he was granted £1,000 by the government at the end of the war, this did not cover his personal expenditure on Colossus, so he was actually out of pocket as a result of his work. Flowers was also awarded the MBE (the same honour later awarded to The Beatles), for work designated simply 'secret and important': no details were given. His career was hamstrung by the fact that he could not reveal anything about his wartime work, and so was unable to persuade his superiors to pursue the development of electronic telephone exchanges in the post-war years. This may sound trivial, but in these days of instant global communication it is hard even for those who were around at the time to remember how primitive telephones were even in the 1950s, when 'long-distance' calls (that is, anything out of town) still had to be connected by a human operator plugging leads into the appropriate sockets. It was ten years after the end of the war before the Post Office began to move into the electronic era, missing out, apart from anything else, on the opportunity to boost British exports at a time of economic hardship. But Flowers lived just long enough to see the importance of his work beginning to be recognized by the computing community. He was able to give a talk in Boston in 1982 which lifted a corner of the veil of secrecy, and in 1997, on the occasion of his own eightieth birthday, Bill Tutte gave a talk detailing the way Tunny was broken. Thomas Flowers died in 1998 at the age of ninety-two.

Unlike Flowers, Alan Turing was able to pick up the threads of his wartime work after the completion of the Delilah project in 1945. He too was 'honoured' by the government, with the award of an OBE – one step up from an MBE, but such an inadequate recognition of his true worth that when Max Newman was also offered an OBE he refused it in protest at Turing's 'ludicrous' treatment.[16]

In October 1945, less than ten years after the publication of 'On Computable Numbers', Turing joined the National Physical Laboratory (NPL) at Teddington, in charge of a project to design and build an electronic 'universal computing machine'. He was, in fact, head-hunted for the post by John Wormersley, the head of the mathematical research division at NPL, who had been an admirer of Turing's work since reading 'On Computable Numbers'. The first fruit of this project was a report by Turing, produced before the end of the year, called 'Proposed Electronic Calculator'. This contained the first full description of a practical stored-program computer – one in which the program is stored in the computer's memory, rather than being plugged in by hand. Each program, remember, can be a virtual machine in its own right, so a single computer can simulate other computers; when you open an app on a tablet or smartphone, you are actually opening a stored program that is itself equivalent to a computer. Turing's plan, as set out in this document, was more far-reaching than the work of his contemporaries in the United States (discussed in the next chapter). He was interested in developing an adaptable machine that could, through its programming, carry out many different tasks; he suggested that one program could modify another; and he understood better than his contemporaries the use of what we now call subroutines. Unlike modern computers, Turing's machine did not have a central processing unit, but worked in a distributed way, with different parts working in parallel with one another; also, instead of the instructions in a program being followed one after another in order, the program (or the programmer!) was to specify which instruction to go to next at each step. All of this made his planned computer faster and more powerful than those planned by his contemporaries; but it would require very skilled programmers to operate it. Why did Turing follow this route? As he wrote to a friend: 'I am more interested in the possibility of producing models of the brain than in the practical applications to computing.'[17]

The trouble was that, as usual, Turing was far ahead of everyone else, and wanted to build an artificial intelligence before anyone had built a really effective electronic calculator. The project, dubbed ACE (for Automatic Computing Engine), was too ambitious, and Turing's strengths did not lie in project management. He wanted Flowers to work with him, but because of the secrecy surrounding Flowers' wartime work could not explain why his presence would be vital; Flowers stayed at Dollis Hill, collaborating with Turing at arm's length, but was soon ordered to concentrate on his proper job. Things stumbled along, with a great deal of testing but very little computer building, until September 1947, when Turing (whose father had died the previous month) left the project, initially for a year's sabbatical in Cambridge, then moving on to Manchester University. But he left a legacy of programs, a kind of software library, prepared in the expectation of the completion of the project; when ACE eventually was built, its immediate success was largely based on this legacy.

While at King's, Turing developed his running to such an extent that he was planning to enter the trials for the Marathon squad in the 1948 Olympic Games; the plan fell through, according to John Turing,[18] when as a result of a bet

'Alan dived into a lake in January, contracted fibrositis, and thereby put himself out of the Wembley Olympics'.

A cut-down version of ACE, called the Ace Pilot Model, or Pilot Ace, was completed at the NPL after Turing left, and ran for the first time on 10 May 1950. It contained a thousand electronic valves, and used just a third of the amount of electronic equipment of contemporary British computers, but ran five times faster than them. The design was adapted and taken up by the English Electric Company as DEUCE, and thirty-three DEUCE machines were built and used commercially in the 1950s and 1960s – the last one was shut down in 1970. The first 'personal' desk-side computer, housed in a cabinet about the size of a tall kitchen refrigerator, was also based on the ACE design. Marketed by the American Bendix Corporation as the G15, it went on sale in 1954. But even by then, the mainstream of computer design was flowing in a different channel, although the idea of a personal computer was an indication of things to come.

In Cambridge, computer development consciously jumped off from the work in the United States which I will discuss in Chapter 2; even the name of the first Cambridge computer, EDSAC (Electronic Delay Storage Automatic Calculator), was deliberately chosen to show its relationship to the American EDVAC. Turing, whose philosophy was to minimize the amount of hardware by maximizing the use of software, described it as 'in the American tradition of solving one's difficulties by means of much equipment rather than by thought'.[19] He was right, but did not appreciate that the average computer user could not think as well as he did, or that the cost of equipment would fall so dramatically, so that today any idiot can use a computer.

The team Turing joined in Manchester was headed by Max Newman, by now Professor of Mathematics; Newman had actually taken unidentifiable bits of a dismantled Colossus to Manchester with him, where some of the pieces were incorporated in the first Manchester computer. It had the distinction of being the first to run a successful program on a stored-program electronic computer. The date was 21 June 1948, and the computer was the 'Manchester Baby', with a random access memory (RAM) equivalent in modern terms to 128 bytes. But it worked. The Baby was the forerunner of the Manchester University Mark I computer, for which Turing developed the programming systems. Audrey Bates, one of the MSc students using the Manchester computer under Turing's supervision in 1948–9, asked if she could go on to work for a PhD; she was told by Newman that he thought it unlikely that anyone would ever get a PhD for working with computers.

Input and output for the Mark I used a system familiar from Bletchley Park days – teleprinter paper tape with a five-bit code punched into it. This kind of tape was still in use for communicating with computers well into the 1960s. As an undergraduate taking a very basic computer course as part of my physics degree I had to prepare programs in this way, before the tapes were taken off to another institution and fed into a computer I never saw (Sussex University didn't have its own computer in those days); the output would be returned a couple of days later as another roll of punched tape (usually with errors caused by the incompetent programming). The Mark I was developed into another commercial machine, the Ferranti Mark I, which in the early 1950s was the most powerful 'supercomputer' around – with a RAM of 1 kilobyte.

It used 3,600 valves, housed in two bays each 17 feet long and 9 feet high, and consumed 25 kilowatts of electricity.

Among Turing's many 'firsts', he now became the first person to program a computer to play musical notes by adjusting the speed of 'beeps' sent to a loudspeaker. Learning of this, Christopher Strachey (a nephew of Lytton Strachey), who had been a pre-war contemporary of Turing at King's and had written a program to play draughts on NPL's ACE machine, came to Manchester and wrote programs that could play 'God Save the King', 'In the Mood' and 'Baa Baa Black Sheep'. In 1951, the results were broadcast by the BBC – on *Children's Hour*.[20] Turing also wrote his letters on the Manchester computer's keyboard, thereby probably becoming the first person to use a word processor.[21]

Turing's principal contribution was typically Turing, and the antithesis of what he called 'the American tradition': a very efficient programming system that was easy for Turing to work with, using binary code, but almost impossible for anyone with a lesser intellect to get to grips with. It was Tony Brooker, who joined the team in October 1951, who worked out how to write programs in a language that looked like algebraic expressions, which were translated automatically into the code the machine understood. This was the first publicly available 'high-level' computer language, the forerunner of things like Fortran and Algol.

By the time Brooker came on the scene, Turing, recently elected a Fellow of the Royal Society, had essentially moved on from artificial computing systems, and was deeply immersed in the puzzle of morphogenesis, the biology of how organisms grow and develop. While DEUCE, EDSAC, the Manchester team and other British projects (one developing from secret work for the Admiralty, another at Birkbeck College in London) developed towards the kind of computers we know today, Turing was again ahead of the pack, puzzling over the way living things are controlled by their 'programming'. His paper 'The Chemical Basis of Morphogenesis', published in 1952, is regarded as being as important in this field as 'On Computable Numbers' is in its own.

But by then time was running out for Turing. Early in 1952 he had a brief homosexual relationship with a nineteen-year-old boy who was then involved in a burglary of Turing's house in Manchester. Turing reported the burglary to the police, naively expecting them to help him; when they found out about the relationship they arrested Turing,[22] who was charged and convicted of the offence that 'being a male person [he had] committed an act of gross indecency with . . . a male person'. His burglarious 'friend' was convicted of the same offence, but seen as Turing's victim and discharged. Turing was put on probation, on condition he took a course of hormone treatment. The court probably thought it was being lenient, but the 'treatment', with the female hormone oestrogen, made him impotent and fat, and, worst of all, affected his ability to think clearly and concentrate.

These events are often linked to his untimely death in 1954, officially determined to be suicide. The situation is in fact rather more complicated than the coroner's verdict suggests. At the time of his death the hormone treatment had been over for a year, and friends describe him as being happy. Work was going well. He had left a 'to do' list for himself at work before going home for the weekend, and rather than a suicide note he left behind ready to post a letter accepting an invitation to a forthcoming event at the Royal Society.

Nothing suggests a suicidal frame of mind. So why the verdict? Well, he did die of cyanide poisoning, and there was a partly eaten apple by his bedside, recalling the couplet from *Snow White*:

> Dip the apple in the brew,
> Let the Sleeping Death seep through.

Bizarrely, though, the apple was never tested to see if it contained cyanide, and Turing had a home laboratory (little more than a glorified cupboard) where, just as in his childhood, he dabbled with chemistry experiments. Some of these involved electroplating using potassium cyanide solution, and police called to the scene reported a strong smell of cyanide (the famous 'bitter almonds' smell) in the room. A jam jar containing cyanide solution was standing, uncovered, on the table in Turing's 'lab'. Perhaps significantly, the ability to smell cyanide diminishes over time, as the concentration of the gas increases, and the post-mortem examination showed that Turing's liver had a low concentration of the poison, not consistent with its having been swallowed.

The simplest explanation is that Turing accidentally inhaled a lethal dose of cyanide just before going to bed. At the other extreme, conspiracy theorists have suggested that as a homosexual who knew too many secrets he was murdered on 'official' orders. Somewhere in between there is the coroner's verdict of suicide. But it is useless to speculate. What matters is that on the night of 7 June 1954, at the age of just forty-one, Alan Turing, the founder of modern computing, died as a result of ingesting cyanide.

CHAPTER TWO

Von Neumann and the Machines

T he American approach to electronic computing, which
Turing derided, was championed by John von Neumann
(usually known as 'Johnny') who, as it happens, was actually
born in Hungary as Neumann János Lajos (in Hungarian the
family name comes first) or 'Jancsi' to his family.

Jancsi

He was born in Budapest on 28 December 1903, the eldest
son of a prosperous Jewish banker. His father had benefited
from the general economic prosperity of Hungary following
the turmoil of the 1860s which had led to the creation of the
Austro-Hungarian dual monarchy out of the old Austrian
Empire, and specifically from the easing of anti-Jewish
restrictions and attitudes in the final decades of the nineteenth century. Originally Neumann Miksa (Max Neumann),
in 1913 Jancsi's father acquired an hereditary title, ostensibly
'for meritorious service in the financial field', although it

helped that he also made a substantial contribution to the state coffers.[1] As a result, he became Margittai Neumann Miksa – in German, Maximilian Neumann von Margitta, Margitta being the ancestral home of the Neumann family and also a play on the name of Max's wife, Margit. So Jancsi became Margittai Neumann János, in German Johann Neumann von Margitta, later simplified to Johann von Neumann, and in English, later still, Johnny von Neumann. As if that were not confusing enough, his two younger brothers, who like Johnny emigrated to the United States, each adopted a different version of the surname. Michael (originally Mihály) simply dropped the 'von' and became Michael Neumann, while Nicholas (originally Miklós) added it to the surname to become Nicholas Vonneumann.

But all that lay far in the future when Jancsi, growing up in Budapest, was first showing signs of his unusual mental ability. He could recall, verbatim, the contents of any book that he read, and devoured the contents of his father's library. Initially the children were educated at home, benefiting from the attentions of both French and German governesses, with specialist tutors in other subjects. During the First World War, they learned English from two British men who had been interned as enemy aliens but were under the protection of Max von Neumann. The family were scarcely touched by the war or its aftermath, including the short-lived but bloody communist regime of 1919. At that time, housing was 're-allocated', ostensibly on the basis of equal shares for all, and a Communist Party official came to assess the needs of the von Neumann family, who had a very large apartment. Nicholas Vonneumann recalled that his father left a bundle of British pound notes on the piano while the assessment was being carried out. After the assessment, the money had gone and the family retained the apartment.[2] When things grew uncomfortable in Budapest, they simply went away to their summer home near Venice for a few weeks. Even so, the experience left Johnny von Neumann, who turned sixteen at the end of 1919, with a lifetime loathing of communism in all its forms.

Jancsi's high school education had begun almost at the same time as the war, in 1914, when he was ten. He attended the Lutheran Gymnasium in Budapest, an elite school famous for its mathematical teaching, where his talent was quickly appreciated and nurtured.[3] By the age of thirteen, he was being fast-tracked not just by the teachers at the school – advanced mathematicians who carried out research alongside their teaching – but also by Joseph Kürschák, a professor at the University of Budapest. At seventeen he published an original paper (co-authored with one of his teachers), and on leaving school that year, 1921, he wanted to become a mathematician. Max von Neumann was horrified, pointing out that there was no money to be made in mathematics. But father and son agreed a compromise – the kind of compromise which made perfect sense to Johnny. He enrolled to study a 'sensible' subject (chemical engineering) at the University of Berlin, later transferring to the Eidgenössische Technische Hochschule (ETH: the Swiss Technical High School) in Zürich, and simultaneously as a mathematics student at the University of Budapest, where he was already well known. Over the four years up to 1925 he duly attended lectures in Berlin and Zürich, and, by arrangement with his professor, popped in to the University of Budapest at the end of each term solely to take examinations. He also prepared a thesis, on set theory, to be offered to the University of Budapest for a doctorate. His final examinations at both institutions were taken in 1925 (he passed, of course, with the highest possible grades) and his doctorate was awarded in 1926. No more was heard of chemical engineering as a career, and von Neumann started his career in mathematics as a *Privatdozent* (the most junior kind of lecturer) at the University of Berlin the same year, while working for a qualification known as the *Habilitation*, a kind of higher doctorate required in the German system before becoming a professor.

One of von Neumann's examiners for his doctorate was the great David Hilbert, the most influential mathematician of his generation (he had been born in 1862), whose work would, as we have seen, soon be the inspiration for Turing's investigations into (in)computable problems. Hilbert was based at the University of Göttingen, and alongside his post in Berlin von Neumann was awarded a Rockefeller Fellowship, which enabled him to carry out research in Göttingen with Hilbert during the academic year 1926/7. Between 1926 and 1929 he published twenty-five scientific papers and established a reputation as a quantum theorist – his book *Mathematical Foundations of Quantum Mechanics*, which was published in 1932, was highly influential, coming to be regarded as a standard text – although, as we shall see, it contained one major flaw. Still in his twenties, he was a glittering star in the mathematical firmament, and his fame had spread worldwide. He had scarcely moved on from Berlin to a more senior post in Hamburg when in 1929 he was invited to visit Princeton to lecture on quantum theory. Replying that he had some personal matters to attend to first, von Neumann made a flying visit to Budapest to get married, to Marietta Kovesi, before becoming a visiting lecturer at Princeton University in February 1930.

Although von Neumann was actually a rather poor lecturer – he had a habit of writing complicated equations in a small corner of the blackboard and erasing them before the students could copy them down – the visit led to his appointment as a professor at Princeton the following year. He loved America, America seemed to love him, and although at first he still held scientific posts in Germany and visited there in the summers, the developing political situation made this a less and less attractive arrangement. Things came to a head in 1933. In January, von Neumann was offered the opportunity to become one of the founding professors at the new Institute for Advanced Study at Princeton (with a starting salary of $10,000), and a few days later Adolf Hitler was appointed Chancellor of Germany. Within months, the new regime began a purge of Jewish academics, and von Neumann eventually resigned from his German positions, although he continued to visit Europe throughout the 1930s.

Johnny and the Institute

The Institute for Advanced Study was founded in 1930, with funding from the philanthropist brother and sister Louis Bamberger and Caroline Bamberger Fuld. They had originally intended to endow a medical school, but after being advised that there were already plenty of medical schools in the United States were persuaded to commit $5 million to set up a research institution where scholars would be free of any diversions such as teaching, and, with life tenure, any worries about the future. Mathematics was the obvious faculty to start the venture with, since at that time mathematicians famously needed nothing more than pencil and paper to work with. The Institute operated on the basis of two terms, running from October to April either side of an extended Christmas break, and required of its academics only that they be in residence during termtime, which amounted to about half the year.

The basis for such an idyllic institution has been questioned, most notably by Richard Feynman, who wrote in his book *Surely You're Joking, Mr. Feynman?*:

> When I was at Princeton in the 1940s I could see what happened to those great minds at the Institute for Advanced Study, who had been specially selected for their tremendous brains and were now given this opportunity to sit in this lovely house by the woods there, with no classes to teach, with no obligations whatsoever. These poor bastards could now sit and think clearly all by themselves, OK? So they don't get any ideas for a while: They have every opportunity to do something, and they're not getting any ideas. I believe that in a situation like this a kind of guilt or depression worms inside of you, and you begin to worry about not getting any ideas. And nothing happens. Still no ideas come. Nothing happens because there's not enough real activity and challenge: You're not in contact with the experimental guys. You don't have to think how to answer questions from the students. Nothing!

But the timing was perfect, as Jewish and other scholars fleeing Nazi persecution began to look for safe havens, and the success of the Institute was assured when Albert Einstein, the most famous of these scientific refugees, was persuaded to make Princeton, rather than Caltech, his base.

Von Neumann, of course, was not a political refugee, but had already chosen to make his career in the United States. Like Einstein, he was a founding member of the IAS's School of Mathematics, which opened in 1933 and was quickly followed by Humanities in 1934 and Economics and Politics in 1935. Johnny stood out from the crowd in many ways. Always smartly dressed (often described as looking 'like a banker'), he had a predilection for large cars, buying a new Cadillac every year, though he was a terrible driver and often got speeding tickets. He loved to hold lavish parties, and according to his daughter, Marina, he needed only three or four hours' sleep a night.[4] Marina was born in 1935, but by the time she was two her parents were divorcing, although they remained on friendly terms and she saw both of them. Johnny had met another Hungarian, Klára (Klári) Dán, on his visits to Europe, and decided they were soulmates. Klári was (by her own admission) a spoiled little rich kid, born in 1911 and by 1937 nearing the end of her second marriage. The situation was complicated not just by the developing political storm (in Europe in 1938 to visit Niels Bohr's institute in Copenhagen, Johnny was struck by the fact that the trains through Germany were full of soldiers) but by the need for getting two divorces, in different countries, and fulfilling the requirements to get Klári an immigration visa for the United States. Johnny had become a US citizen in 1937, but initially retained his Hungarian citizenship as well; unfortunately, Hungarian law (to which he was subject as a Hungarian citizen) did not recognize his American divorce. So he had to renounce his Hungarian citizenship and marry as an American citizen, which in turn enabled Klári to obtain the necessary visa as his wife. The formalities took months, and it was 18 November 1938 before the couple were finally married in Budapest. The following month they sailed for New York from Southampton on the *Queen Mary*, arriving a week before Christmas. The marriage lasted for the rest of Johnny's life, although it did not always run smoothly, and many personal insights into his character come from a memoir Klári wrote, now in the possession of Johnny's daughter Marina. One intriguing titbit is that Johnny, like so many gifted mathematicians, suffered from a mild compulsive disorder, one manifestation of which was that whenever he turned a light on or off he had to do so by flipping the switch exactly seven times.

The couple were childless. Klári suffered a miscarriage in the summer of 1942, and with Johnny often away on secret war work, in 1943 she took a job with the Office of Population Research at Princeton University. The training she gained there in statistical work would prove invaluable when she later worked with Johnny as one of the first computer programmers – but that is getting ahead of our story.

In spite of the distractions of his personal life, and the parties (which continued with Klári as hostess), Johnny was one of the early stars of the Institute, proof that Feynman's reservations did not apply to everyone there. Between 1933 and the end of 1941, when America was forced into war, he published thirty-six scientific papers and, even more significantly, met Alan Turing and learned about his work. But long before the Japanese attack on Pearl Harbor, Johnny was becoming involved with the military. He became a US citizen on 8 January 1937, and was so concerned about the situation in Europe that he soon applied for a commission in the reserve of the US Army, being turned down solely on grounds of age (he had just passed the cut-off age of 35 by the time he had completed the required examinations). Instead, he became a consultant to the US Army's Ordnance Department, and then (in 1940) a member of the scientific advisory board for its Ballistic Research Laboratory. Other appointments followed, including membership of the National Defense Research Council (NDRC) and work for the US Navy (which, among other things, took him on an extended visit to Britain). By the time the Japanese bombs fell on Pearl Harbor on 7 December 1941, he was the go-to man for advice about large explosions, had a reputation for being able to break down complex problems into their simple components, and was a skilful chairman of committees who got things done. Inevitably – although he kept fingers in many pies – he was drawn towards the Manhattan Project.

Johnny and the Bomb

In July 1946, von Neumann received both the US Navy's Distinguished Civilian Service Award and President Harry S. Truman's Medal for Merit. The citation for the latter said that he was 'primarily responsible for fundamental research by the United States Navy on the effective use of high explosives, which has resulted in the discovery of a new ordnance principle for offensive action'. The reference was, of course, to the nuclear bomb.

Von Neumann had been called back from England to join the Manhattan Project in 1943, and by the autumn he was in Los Alamos, where he made two key contributions to the project.[5] The first was to point out (and prove mathematically) that such a bomb would be more effective if exploded at altitude rather than at ground level, because both the heat and blast would affect a wider area. The second contribution was much more profound.

A fission bomb works by bringing together forcefully a sufficient amount ('critical mass') of a radioactive material such as uranium or plutonium. Under such conditions, particles (neutrons) released by the fission ('splitting') of one atomic nucleus trigger the fission of more nuclei, in a runaway chain reaction. Each 'split' converts a little mass into energy, in line with Einstein's famous equation, and the overall result is the explosive release of a lot of energy as heat, light, and other electromagnetic radiation. But if the critical mass is not tightly confined, most of the neutrons escape and the material simply gets hot, rather than exploding. The first method the Los Alamos team considered for achieving the required result was to prepare a critical mass of uranium in two halves at opposite ends of a tube, and fire conventional explosives to smash one of them (the 'bullet') into the other. For obvious reasons, it was called the gun method, and was used in the Little Boy weapon dropped on Hiroshima. But this method could not be used with plutonium for technical reasons, not least the possibility that the plutonium, being more active than uranium, might 'pre-ignite' just before the bullet hit the target, releasing its neutrons too gradually and causing the bomb to fizzle rather than explode. This was where von Neumann came in.

Before von Neumann arrived in Los Alamos, another member of the team, Seth Neddermeyer, had suggested setting off explosives to produce shock waves which would squeeze a 'subcritical' mass of plutonium to the point where it reached critical mass and exploded. But the idea had not been followed up, and nobody had been able to work out exactly how to achieve the desired result. Edward Teller, a member of the Los Alamos team who is remembered as the 'father of the hydrogen bomb', later recalled how the problem was solved.[6] Von Neumann had calculated the kind of pressure that would be produced in a lump of plutonium if it was squeezed in the grip of a large number of explosive charges surrounding it and going off simultaneously. He discussed his results, which still seemed to fall short of what was necessary to produce a practical bomb, with Teller, who had worked in geophysics and knew that under very high pressures such as those at the centre of the Earth even a substance like iron gets compressed to higher density than at the planet's surface. He pointed out that this compressibility would make the process von Neumann was describing even more effective, because the more the plutonium atoms were squeezed together the easier it would be for a chain reaction to take place. Von Neumann redid the calculation, taking account of compressibility, and found that the trick would work. After a great deal more work by many people, including von Neumann, the result was the Fat Man bomb dropped on Nagasaki, in which a hollow shell of plutonium was triggered into explosive fission by the firing of thirty-two opposing pairs of detonators to produce an inward compression. This process owed a great deal, in the days before electronic calculators, to von Neumann's ability at carrying out mathematical calculations; but it also highlighted the need for faster methods of carrying out such calculations, which could be used when there wasn't a von Neumann around to do the work. This became of crucial importance when von Neumann, who continued to spend two months each year visiting Los Alamos after the war, became involved in the development of the hydrogen bomb, based on nuclear fusion, not fission: because even von Neumann couldn't do all the calculations on his own.

The calculations for the Manhattan Project had been aided by the use of machines, supplied by the company International Business Machines (IBM), which correlated data using sets of punched cards. These were in no sense computers in the modern meaning of the word, but moronic machines, using mechanical switches, that could be set up to perform basic arithmetical operations. For example, a machine could take two cards punched with holes corresponding to two numbers (say, 7 and 8) and add them up, spitting out a card punched with holes corresponding, in this case, to the number 15. They could alternatively be set up to carry out subtraction or multiplication, and they could handle large numbers of cards. Several machines could be set up so that the output from one machine became the input of the next, and so on. In this way they could carry out tasks along the lines of 'take a number, double it, square the result, take away the number you first thought of and punch a card with the answer on'.

In *Surely You're Joking, Mr. Feynman?*, Richard Feynman describes how machines like this were used to carry out the donkey work of computations for the Manhattan Project. It was a colleague, Stanley Frankel, who realized the potential of the IBM machines, but Feynman ended up being in charge of their operation. The first step was to break down the calculations involved in working out things like the compressibility of plutonium into their individual components – something now familiar to anyone who does computer programming. The instructions for the complex calculations were then passed to a team of 'girls' (as Feynman describes them), each armed with a mechanical calculator, like a glorified adding machine, operated by hand. One girl did nothing but multiply the numbers she was given and pass the result on to the next girl; one did nothing but calculate cubes; and so on. The system was tested in this way and 'debugged', in computer jargon, until Frankel and Feynman knew it worked, then put on to a production line basis using the IBM machines. In fact, after a bit of practice the team of girls was just as fast as the room full of IBM machines. 'The only difference,' says Feynman, 'is that the IBM machines didn't get tired and could work three shifts. But the girls got tired after a while.' Von Neumann was closely involved with this project, as one of Feynman's 'customers', and learned all about the operation of the system in the spring of 1944. To all intents and purposes, this was computing without a computer, with Feynman as the programmer, and it highlights the point that for all Turing's hopes we do not yet have anything like a mechanical intelligence; we only have machines that can do the same thing as a team of humans, but faster and without tiring. In either case, the team, or the machine, needs a human programmer.

The American heritage

Computers of the kind we use today owe as much to von Neumann as to Turing, and in his post-war work von Neumann built on a heritage of American developments. The prehistory of electronic computing in the United States had two strands, one involving computing and the other involving electronics. The direct line to the IBM card-sorting machines which Feynman used to carry out calculations for von

Neumann goes back to the American census of 1890. The tenth US census of 1880 (and all previous ones) had been tabulated by hand, with hundreds of clerks copying stacks of information from the record sheets into various categories. But with the US population growing so rapidly through immigration, the point was being reached where it would be impossible to tabulate the results of one census fully before the next census was due. John Billings, who was in charge of statistical analysis for both the 1880 and the 1890 censuses, was well aware of the problem, and early in the 1880s this led him into a conversation recalled by his colleague Herman Hollerith in 1919:

> One Sunday evening at Dr. Billings' tea table, he said to me there ought to be a machine for doing the purely mechanical work of tabulating population and similar statistics. We talked the matter over and I remember [that] he thought of using cards with the descriptions of the individual shown by notches punched in the edge of the card. [I] said that I thought I could work out a solution for the problem and asked him if he would go in with me. The Doctor said he was not interested any further than to see some solution of the problem worked out.[7]

So it was Hollerith who put flesh on the bones of Billings' idea, and who by the time of the 1890 census had developed a system based on punched cards (which he chose to be the size of a dollar bill) where the pattern of holes punched in the cards indicated characteristics of the individual, such as whether they were born in the United States or not, their sex, whether they were married or not, how many children they had and so on. The cards were read by an electromechanical sorter which could, for example, take a batch of cards from a particular city and select from them all the ones corresponding to, say, married white men born in the United States with at least two children.

The success of the equipment, used in the 1890 census to process the records of some 63 million people, using 56 million cards, enabled Hollerith to set up the Tabulating Machine Company in 1896. This evolved into the Computer-Tabulating-Recording Company in 1911, and then into the International Business Machines Corporation (IBM) in 1924. By 1936, when Turing published 'On Computable Numbers', the world was using 4 billion cards each year; with hindsight, each card could be regarded as a 'cell' in the endless tape of a Turing machine. The reason for this growth in the use of punched cards was that it had been realized that they could be used not just to tabulate statistics, but to carry out arithmetical operations. And the need to mechanize arithmetic had been made clear by the requirements of the military in the First World War, just as the need for high-speed electronic computers would be made clear by the requirements of the military in the Second World War.

The specific military requirement that encouraged the development of punched-card computing was the need to calculate the flight of shells fired from guns, and later the fall of bombs dropped from aircraft. Ballistics would be a simple science on an airless planet, where projectiles would follow parabolic trajectories described beautifully by Newton's laws, under the influence of gravity. Unfortunately, in the real world shells in flight are affected by the density of the air, which changes with altitude, temperature, humidity and other factors, as well as the initial velocity of the projectile. In order to hit a given target, even assuming there is no wind to deflect the shell in its flight and ignoring the subtleties caused by the rotation of the Earth, the gun has to be elevated at an angle which takes all of these factors into account. As if this were not tricky enough, each gun has its own firing characteristics; before guns could be supplied to the army they had to be tested, with a so-called 'firing table' being worked out for each individual weapon. In the field, the gunners would have to refer to these tables in order to determine exactly how to elevate their guns under different conditions. A typical firing table had several thousand entries, corresponding to different trajectories, and it would take several hours for a human armed with a desk calculator to work out a single trajectory. The result was a major bottleneck between the manufacture of guns in factories and their delivery to armies in the field.

One of the leaders in the field of military ballistics in the United States in the First World War was the mathematician Oswald Veblen, who headed a team at the Army's Aberdeen Proving Ground in Maryland. He would later, as a professor at Princeton University, play a major role in establishing the Institute for Advanced Study, and in particular in ensuring that it started life with a group of eminent mathematicians, including Johnny von Neumann.

In the decade following the war, Hollerith-type punched-card machines began to be used for some scientific purposes, especially the tedious calculation of astronomical tables, and there were parallel developments which have been described by Herman Goldstine, who became von Neumann's right-hand man, in his book *The Computer from Pascal to von Neumann*. But here I shall focus on the thread that led directly to von Neumann himself. A key step came in 1933, when the

Ordnance Department of the US Army established a joint venture with the Moore School of the University of Maryland to develop an improved calculating machine. Another involved John Presper Eckert, working at Columbia University on problems of celestial mechanics, originally using standard IBM machines of the late 1920s; the success of this work encouraged IBM to produce a special machine, called the Difference Tabulator, for the Columbia astronomers. Out of this collaboration grew the Thomas J. Watson Astronomical Computing Bureau,[8] a joint venture of the American Astronomical Society, Columbia University and IBM. This, says Goldstine, 'marked the first step in the movement of IBM out of the punch card machine business and into the modern field of electronic computers'. IBM's interest was stimulated further by Howard Aiken, of Harvard University, who proposed developing a system based on punched-card machines to produce an electromechanical computer; a project based on his ideas began in 1939 and achieved success in 1944, although by then it was being overtaken by developments in electronic computing. Another electromechanical device, developed at Bell Laboratories and using telephone switching relays, was also completed in 1944, and suffered the same fate. But the Americans were not the first to do this.

A German diversion

In the mid-1930s, a German engineer called Konrad Zuse, working in the aircraft industry, developed an electromechanical calculating machine using binary arithmetic. He was completely ignorant of developments in other parts of the world, and worked everything out from scratch. This Z1, completed in 1938, used on/off switches in the form of steel pins that could be positioned to the right or left of a steel lug using electromagnets. A friend of Zuse's, Helmut Schreyer, suggested that vacuum tubes (of which more below) would be more efficient, and the two of them calculated that a machine using 2,000 tubes would be feasible; but, like their counterparts in the United States they felt that the technology of the time was too unreliable to be put to practical use immediately. With the outbreak of war, Zuse was called up for the army, but after six months he was discharged to work at the Henschel aircraft factory, where he was involved in the development of the V1, the first cruise missile. He offered the authorities his and Schreyer's idea of a 2,000-tube computer to use in directing anti-aircraft fire, but when he said the project would need two years to come to fruition he was told it was not worth funding because the war would be over by then.

Instead, working essentially alone, in his spare time, with some help from friends, under difficult wartime conditions, Zuse developed a new machine, using electromechanical relays of the kind used in telephone systems. The complete machine was known as Z3, because Z2 was assigned to a smaller machine used to test some of the components. Z3 was intended to be the electromechanical equivalent of an electronic tube machine, so that the switch to tubes could be made easily as the next step, with tubes replacing relays. Z3 was a programmable machine with 2,400 relays, 1,800 used for memory and 600 in the calculating part of the machine. It was completed in 1941 – the world's first programmable electromechanical digital computer, but not the first electronic computer, and not the first stored-program computer.[9] The success of the Z3 and the fact that the war showed no sign of ending encouraged support for a more powerful machine, to be called Z4; although this was not completed before the end of the conflict, it was the only one of Zuse's machines to survive the bombing, although replicas of his Z1 and Z3 machines have since been built. Zuse was able to smuggle the partly built Z4 out of Germany to the ETH in Zürich, where it was completed in 1950 (making it arguably the world's first operational commercial computer) and ran until 1955. By the time Zuse was able to pick up the threads of his work, however, he had been overtaken by events elsewhere. His story is no more than a detour on our route to the modern machine.

The second strand

Electronics, the second strand of the American heritage, had developed rapidly in the 1920s and 1930s. At the beginning of the twentieth century, John Fleming in Britain and Lee de Forest in the United States had independently invented the thermionic valve, known in the US as the vacuum tube. Each name describes part of the function of the device. In an evacuated glass tube, a stream of electrons flows from a cathode through the vacuum to be picked up by an anode. The flow of electrons can be switched on and off by applying a secondary electric current to a 'grid' in the tube, just as the flow of water in a pipe can be controlled by a valve. The tube acts like a valve controlling the flow of electrons. Later refinements allow for more sophisticated interactions with the electron beam, but in the context of the development of computers, it is the fact that they can be on or off, corresponding to the binary states 0 or 1, that matters.

One big incentive encouraging the development of

electronic valves was (or perhaps I should say, given Flowers' experience, should have been) the need for automatic telephone exchanges; the other was the prospect of television. Like the valve/tube, television was developed more or less simultaneously in the UK and US. Many people contributed, but the relevant thread here can be traced through the work of Vladimir Zworykin, a Russian-born inventor who studied with Boris Rosing in St Petersburg and Paul Langevin in Paris before becoming a signals officer in the Russian Army in the First World War. In 1918, aged twenty-nine, during the turmoil of the Russian Civil War Zworykin fled to America; he returned to work for the White government, but when the Reds won the conflict, he settled in the US. There he worked for Westinghouse in Pittsburgh and then for RCA, developing a camera tube system called the iconoscope, not unlike the cathode ray tubes used by home televisions before the advent of 'flat' screens. Zworykin's cathode ray tubes were used both as transmitter and as receiver.

In an interview reported by Albert Abramson, Zworykin said that the development of electronics in the first half of the twentieth century occurred in three stages.

> In the first, beginning with [de Forest] in 1906 and ending with the First World War, electron currents were controlled in vacuum tubes in much the same manner as a steam valve controls the flow of steam in a pipe . . . no more attention was paid to the behavior of the individual electrons in the tube than is customarily expended on the motion of the individual steam molecules in the valve.

In the second stage, in the 1920s, 'the directed, rather than random, character of electron motion in vacuum was applied in the cathode-ray tube'. And in the third stage, in the 1930s and later, beams of electrons were divided into groups 'either on the basis of time, the electrons being bunched at certain phases of an applied high-frequency field,[10] . . . or of space, as in image-forming devices'.[11]

So the third stage of the development of electronics in the US was well under way by the time gathering war clouds in Europe spurred the development of faster calculating devices, both for working out conventional firing tables and for the even trickier problem of improving the success of anti-aircraft gunnery by calculating in real time the speed and altitude of aircraft in order to set fuses to explode at the right height. This ambitious objective would not be achieved in time to influence anti-aircraft gunnery during the Second World War. But the spinoffs from making the attempt were immense.

Almost as soon as Germany invaded Poland in 1939 the US army commissioned RCA to begin work on such a computer. Jan Rajchman, a Polish immigrant working for Zworykin's team, led the way in developing a technique for sorting and switching pulses of electrons inside vacuum tubes in such a way that a single tube could be used to multiply two numbers together and add a third number to the answer, working in binary maths. The result was a single-purpose machine called the Computron; it's some indication of how difficult this work was, using 1940s technology, that a patent application for such a 'Calculating Device' was not filed until July 1943. By then, the RCA team was also working to develop a data storage device – what we would now call Random Access Memory (RAM) – using vacuum tubes; it would be known as the Selectron, but would prove rather unwieldy in practice, and was overtaken by other developments. All of this activity caught the attention of von Neumann, who was a frequent visitor to RCA, and he decided that this was the 'go-to' place for the development of electronic digital computers. By then, too, he had something to build on.

ENIAC

Herman Goldstine had obtained a PhD in mathematics from the University of Chicago in 1936, and stayed there for the next three years, working with the mathematician Gilbert Bliss and becoming known to, among others, Oswald Veblen. He was teaching ballistics at the University of Michigan when, in the summer of 1942, he was called up to serve with the Army Air Force and sent to California for training before being posted overseas. Aware that there were better uses for Goldstine's talents, Oswald Veblen officially requested that he be posted instead to the Aberdeen Proving Ground. Goldstine actually received two sets of orders on the same day, one instructing him to ship out across the Pacific, the other ordering him to report to Aberdeen. He set off east as fast as he could, with fresh lieutenant's bars on his shoulders.

Goldstine was assigned to the Moore School, where it was becoming clear that however many human 'computers' were employed they would never be able to clear the firing table bottleneck. There, in the autumn of 1942, he met John Mauchly, a physicist who had been teaching at Ursinus College in Philadelphia at the outbreak of war; recruited to a training course in electronics at the Moore School, he had shown such aptitude that he was asked to join the faculty before he had completed the course. 'By August 1942,' writes Goldstine, he had advanced his thinking [about computing machines] enough to write a brief memorandum summarising his ideas; this was circulated among his colleagues and perhaps most notably to a young graduate student, J. Presper Eckert, Jr., who was undoubtedly the best electronic engineer in the Moore school. He immediately, as was his wont, immersed himself in the meager literature on counting circuits and rapidly became an expert in the field.

Mauchly took the lead on a proposal, pushed forward by Goldstine and received with enthusiasm by Veblen, for a digital electronic computer; submitted on 2 April 1943, the proposal was accepted just a week later. Eckert became the chief engineer on the project, which developed at frantic wartime pace. The end product, little more than two and a half years later, was a machine known as ENIAC (for Electronic Numerical Integrator And Computer), which contained 17,468 vacuum tubes, about 70,000 resistors and 1,500 relays, weighed 30 tons and used 174 kilowatts of electricity. It stood 10 feet high and 3 feet wide, and the cabinets, which stood side by side in a large 'U' shape, would have stretched out along a single line for 100 feet if laid end to end. Input and output used IBM card readers and punches.

Right at the start of the project, Goldstine and his engineer colleagues made visits to the RCA Research Laboratories in Princeton, becoming familiar with the work of Zworykin and Rajchman. The contract for the construction of the machine, signed with the Moore School on 5 June 1943, had originally been offered to RCA, but Zworykin, in a rare moment of under-optimism, expected it to fail and turned the project down, although RCA staff, including Jan Rajchman, were still closely involved as consultants. Designed for the specific job of preparing firing tables, ENIAC was flexible enough to be adapted to a stored program system after the war, and ran until October 1955, tackling problems that included weather prediction, cosmic-ray studies and wind-tunnel design; but in its original incarnation it was 'programmed' using a plugboard system that could take weeks to set up for a specific type of problem. The conversion, based on a 51-word instruction set devised by Herman Goldstine's wife Adele, 'slowed down the machine's operation, [but] speeded up the programmer's task enormously . . . the old method was never used again'.[12]

A friend of mine, Lawrence Lerner, recalls working at the Aberdeen Proving Ground in 1953, writing programs in machine language for ENIAC, which 'was very impressive to look at, mostly through windows from the control room'. At that time, both the relay computer and ENIAC were still functioning. The relay machine, he says,

> was more fun to watch; it filled a large room with glass-topped cases full of relays, and there were seemingly random clicks from all over the room as the machine executed its program. But ENIAC was nice to work with because it was air conditioned. This was to keep the vacuum tubes cool, but it was nice for the people, too.

Lawrence also remembers attending a seminar in the summer of 1953 where 'an engineer demonstrated a circuit that could translate decimal numbers into binary; it was the death-knell of binary machine-language programming' and a key step towards the modern user-friendly computers. But ENIAC was not, as has sometimes been claimed, the world's first 'proper' stored-program electronic computer. As we have seen, that honour goes to the Manchester Baby, while on 6 May 1949 Cambridge University's EDSAC 1 (calculating the squares of the numbers from 0 to 99) became the first complete and fully operational electronic digital stored program computer. Nor was ENIAC the world's first electronic computer; that honour, of course, goes to Colossus.

Von Neumann visited the ENIAC project for the first time in the late summer of 1944. The visit came about as a result of a chance encounter. Goldstine was waiting for a train when he saw von Neumann standing alone on the same platform. Although they had never met, he approached the great man and began to talk to him. Von Neumann, he says, was

> a warm, friendly person who did his best to make people feel relaxed in his presence [but when] it became clear to von Neumann that I was concerned with the development of an electronic computer capable of 333 multiplications per second, the whole atmosphere of our conversation changed from one of relaxed good humor to one more like the oral examination for the doctor's degree in mathematics.

That chance meeting changed the path of von Neumann's career, and the course of the development of computing.

Von Neumann picks up the ball

ENIAC was officially dedicated at a ceremony in February 1946, when it computed the flight of a shell in twenty seconds – two-thirds of the flight time of the hypothetical shell itself. Impressive stuff by pre-war standards, but too late to affect the outcome of the war. What was not publicly revealed at the time, though, was that since December 1945 the machine had actually been working on a much more important problem – one that would affect the development of the next (Cold) war. This secret operation of the machine overlapped with testing, blurring the line so that it is impossible to point to a specific day as the moment ENIAC became operational; but it was certainly up and running by 1 January 1946, and arguably by 1 December 1945.

Von Neumann's interest in ENIAC stemmed in no small measure from his involvement with the Manhattan Project. By the time ENIAC was ready for testing, the fission (atomic) bomb was a reality and some scientists, including Edward Teller and von Neumann, were pressing for the development of far more powerful fusion (hydrogen) bombs. The controversial political decision to go ahead with the development of what was known at the time as the 'super' bomb was made out of fear of Soviet intentions following the defeat of Hitler's Germany. But designing such a bomb would require much more computation than the design of the fission bombs. Von Neumann was uniquely placed – the only person who was privy to both the secrets of Los Alamos and developments at the Moore School. He was also the only person with the prestige and influence to ensure that the first program actually run on the ENIAC, starting in December 1945 before the machine was used for its intended purpose, was a simulation for the super project. In a striking example of pragmatism, the ENIAC team were allowed to see the equations, which were not classified, without being told anything about the super bomb, which was. The Los Alamos 'problem' was transferred to the Moore School in the form of a million IBM/Hollerith punched cards.[13] The simulation worked, as far as the maths went, although in the end it proved useless because the assumptions about the physics built into it turned out to be wrong. But it worked by using the brute force approach to number crunching that Turing abhorred. The pattern was already being established. Before ENIAC was complete, its designers and engineers were planning a new machine, to be called EDVAC (Electronic Discrete Variable Automatic Computer). It was at this point that von Neumann picked up the ball and more or less ran off with it, in the process setting a precedent for computer design that would last for decades, and is still influential today.

Von Neumann, with his contacts in Los Alamos, Washington, Aberdeen, Princeton and Philadelphia, was the catalyst who brought the various threads together and persuaded the authorities to continue funding computer developments after the war. But plans for the kind of machine to succeed ENIAC were drawn up by a large team at the Moore School, with Mauchly and Eckert leading the discussions and contributions from Goldstine, his wife Adele, and Arthur Burks. Von Neumann kept in touch largely by letter, corresponding with Goldstine and keeping up to date with thinking at the Moore School while making suggestions from afar. In the planned EDVAC, the various components of the computer would be separated into different units – a central processing unit (CPU) to do arithmetic, a memory, some sort of control system, and input/output devices – with data being shuffled between them as required. There would also be a fifth, less visible, component: a means of communication between the other four units, in particular the memory and the CPU, called a bus. A key feature of this kind of computer architecture is that problems are solved serially, step by step, going through a chain of instructions in sequence. 'The notion of serial operation,' says Goldstine, 'was pushed to its absolute limit in the EDVAC design.' In the alternative parallel (or distributed) architecture, favoured by Turing, different pieces of the problem are tackled by different parts of the machine simultaneously (in parallel). To take a simple example, if two long numbers are to be added together, a serial machine will add each digit in turn, just as we do, but a parallel machine will add all the digits at once. The serial method is slower, but needs less hardware. The parallel approach, however, avoids the problem with the serial structure that parts of the machine may be sitting idle, doing the electronic equivalent of twiddling their thumbs, while waiting for another part of the machine to finish a task. A related problem is that with such a structure there are delays caused by the constant need to shuttle data and instructions between the memory and the processor, along the bus, a deficiency which became known as 'the von Neumann bottle-neck'. It got this name because the Moore School plan became known as the von Neumann architecture. And that happened in somewhat acrimonious circumstances.

Working at von Neumann's direction, Goldstine took von Neumann's notes and the letters they had exchanged and prepared a document just over 100 pages long, called 'First Draft of a Report on the EDVAC', which was reproduced in limited numbers and distributed to a select few at the end of June 1945. The ideas discussed in this report were less advanced than those of Alan Turing discussed in the previous chapter, but were much more influential, because of the way they were promoted.

But the way they were promoted did not please everyone.

The snag was that the report appeared with only one name on it as author: that of John von Neumann. Adding injury to insult, the document was later regarded as a publication in the legal sense of the term, placing the ideas in the public domain and preventing them from being patented. And compounding the injury, it transpired that in May 1945 von Neumann had signed a lucrative consultancy deal with IBM, in exchange for which he assigned (with a few exceptions) all the rights to his ideas and inventions to them. Hardly surprising that Eckert later complained that 'he sold all our ideas through the back door to IBM'.[14]

Von Neumann now decided that he really wanted a computer at the IAS, where he was based, and with breathtaking insouciance asked the Moore School team to join him there. Goldstine took up the offer; Mauchly and Eckert left the academic world and formed the Electronic Control Company, which became a pioneering and successful commercial computer business, achieving particular success with their UNIVAC machine (Universal Automatic Computer). The EDVAC project staggered on without them, but by the time it was completed, in 1951, it was largely outdated, made obsolescent by other machines built using the architecture described in the 'First Draft' – machines conforming to what was now widely known as the von Neumann architecture, bottleneck and all.

The scene was now set for decades of development of those ideas, with valves giving way to transistors and chips, machines getting smaller, faster and more widely available, but with no essential change in the logical structure of computers. The Turing machine in your pocket owes as much to von Neumann (who always acknowledged his debt to 'On

Computable Numbers') as to Turing, but is no more advanced in terms of its logical structure than EDVAC itself.

There is no need to go into details about the development of faster and more powerful computers in the second half of the twentieth century. But I cannot resist mentioning one aspect of that story. In a book published as late as 1972, Goldstine commented: 'It is however remarkable that Great Britain had such vitality that it could immediately after the war embark on so many well-conceived and well-executed projects in the computer field.'[15] Goldstine had been at the heart of the development of electronic computers, but the veil (actually, more like an iron curtain) of secrecy surrounding the British code-breaking activities was such that a quarter of a century after the developments described here, he was unaware of the existence of Colossus, and thought that the British had had to start from scratch on the basis of the 'First Draft'. What they had really done, as we have seen, was even more remarkable.

Self-replicating robots

This isn't quite the end of the story of von Neumann and the machines. Like Turing, von Neumann was fascinated by the idea of artificial intelligence, although he had a different perspective on the rise of the robot. But unlike Turing, he lived long enough (just) to begin to flesh out those ideas.

There were two parts to von Neumann's later work. He was interested in the way that a complex system like the brain can operate effectively even though it is made up of fallible individual components, neurons. In the early computers (and many even today), if one component, such as a vacuum tube, failed the whole thing would grind to a halt. Yet in the human brain it is possible for the 'hardware' to suffer massive injuries and continue to function adequately, if not quite in the same way as before. And he was also interested in the problem of reproduction. Jumping off from Turing's idea of a computer that could mimic the behaviour of any other computer,[16] he suggested, first, that there ought to be machines that could make copies of themselves and, secondly, that there could be a kind of universal replicating machine that could make copies of itself and also of any other machine. Both kinds of mimic, or copying machine, come under the general heading 'automata'.

Von Neumann's interests in working out how workable devices can be made from parts prone to malfunction, and in how complex a system would have to be in order to reproduce itself, both began to grow in 1947. This was partly because he was moving on from the development of computers like the one then being built at the IAS and other offspring of EDVAC, but also because he became involved in the pressing problem for the US air force in the early 1950s of how to develop missiles controlled by 'automata' that would function perfectly, if only during the brief flight time of the rocket.

Von Neumann came up with two theoretical solutions to the problem of building near-infallible computing machines out of fallible, but reasonably accurate, components. The first is to set up each component in triplicate, with a means to compare automatically the outputs of the three subunits. If all three results, or any two results, agree, the computation proceeds to the next step, but if none of the subunits agrees with any other, the computation stops. This 'majority voting' system works pretty well if the chance of any individual subunit making a mistake is small enough. It is even better if the number of 'voters' for each step in the calculation is increased to five, seven, or even more. But this has to be done for every step of the computation (not just every 'neuron'), vastly (indeed, exponentially) increasing the amount of material required. The second technique involves replacing single lines for input and output by bundles containing large numbers of lines – so-called multiplexing. The data bit (say, 1) from the bundle would only be accepted if a certain proportion of the lines agreed that it was correct. This involves complications which I will not go into here;[17] the important point is that although neither technique is practicable, von Neumann proved that it is possible to build reliable machines, even brains, from unreliable components.

As early as 1948, von Neumann was lecturing on the problem of reproduction to a small group at Princeton.[18] The biological aspects of the puzzle were very much in the air at the time, with several teams of researchers looking for the mechanism by which genetic material is copied and passed from one generation to the next; it would not be until 1952 that the structure of DNA was determined. And it is worth remembering that von Neumann trained as a chemical engineer, so he understood the subtleties of complex chemical interactions. So it is no surprise that von Neumann says that the copying mechanism performs 'the fundamental act of reproduction, the duplication of the genetic material'. The surprise is that he says this in the context of self-reproducing automata. It was around this time that he also surmised that up to a certain level of complexity automata would only be able to produce less complicated offspring, while above this level not only would they be able to reproduce themselves, but 'syntheses of automata can proceed in such a manner that each automaton will produce other automata which are more complex and of higher potentialities than itself'. He made the analogy with the evolution of living organisms, pointing out that 'today's organisms are phylogenetically descended from others which were vastly simpler'. How did the process begin? Strikingly, von Neumann pointed out that even if the odds are against the existence of beings like ourselves, self-reproduction only has to happen once to produce (given time and evolution) an ecosystem as complex as that on Earth. 'The operations of probability somehow leave a loophole at this point, and it is by the process of self-reproduction that they are pierced.'

By the early 1950s, von Neumann was working on the practicalities of a cellular model of automata. The basic idea is that an individual component, or cell, is surrounded by other cells, and interacts with its immediate neighbours. Those interactions, following certain rules, determine whether the cell reproduces, dies, or does nothing. At first, von Neumann thought three-dimensionally. Goldstine:

> [He] bought the largest box of 'Tinker Toys' to be had. I recall with glee his putting together these pieces to build up his cells. He discussed this work with [Julian] Bigelow and me, and we were able to indicate to him how the model could be achieved two-dimensionally. He thereupon gave his toys to Oskar Morgenstern's little boy Karl.

The two-dimensional version of von Neumann's model of cellular automata can be as simple as a sheet of graph paper on which squares are filled in with a pencil, or rubbed out, according to the rules of the model. But it is also now widely available in different forms that run on computers, and is sometimes known as the 'game of life'. With a few simple rules, groups of cells can be set up that perform various actions familiar in living organisms. Some just grow, spreading as more cells grow around the periphery; others pulsate, growing to a certain size, dying back and growing again; others move, as new cells are added on one side and other cells die on the opposite side; and some produce offspring, groups of cells that detach from the main body and set off on their own. In his discussion of such systems, von Neumann also mentioned the possibility of arbitrary changes in the functioning of a cell, equivalent to mutations in living organisms.

Von Neumann did not live long enough to develop these ideas fully. He died of cancer on 28 February 1957, at the age of fifty-three. But he left us with the idea of a 'universal constructor', a development of Turing's idea of a universal computer – a machine which could make copies of itself and of any other machine: that is, a self-reproducing robot. Such devices are now known as von Neumann machines, and they are relevant to one of the greatest puzzles of our, or any other time – is there intelligent life elsewhere in the Universe? One form of a von Neumann machine would be a space-travelling robot that could move between the stars, stopping off whenever it found a planetary system to explore it and build copies of itself to speed up the exploration while sending other copies off to other stars. Starting with just one such machine, and travelling at speeds well within the speed of light limit, it would be possible to explore every planet in our home Milky Way galaxy in a few million years, an eyeblink as astronomical timescales go. The question posed by Enrico Fermi (Why, if there are alien civilizations out there, haven't they visited us?) then strikes with full force.

There's one other way to spread intelligence across the Universe, of which von Neumann was also aware. A universal constructor would operate by having blueprints, in the form of digitally coded instructions, which we might as well call programs, telling it how to build different kinds of machines. It would be far more efficient to spread this information across the Universe in the form of a radio signal travelling at the speed of light than in a von Neumann machine pottering along more slowly between the stars. If a civilization like ours detected such a signal, it would surely be copied and analysed on the most advanced computers available, ideal hosts for the program to come alive and take over the operation of the computer. In mentioning this possibility, George Dyson makes an analogy with the way a virus takes over a host cell; he seems not to be aware of the entertaining variation on this theme discussed back in 1961 by astrophysicist Fred Hoyle in his fictional work *A for Andromeda*,[19] where the interstellar signal provides the instructions for making (or growing) a human body with the mind of the machine. Hoyle, though, was well aware of the work of Turing and von Neumann.

There is something even more significant that Turing and von Neumann left us to ponder. How does our kind of intelligence 'work' in the first place? Each of them was convinced that an essential feature of the human kind of intelligence is the capacity for error. In a lecture he gave in February 1947, Turing said:

> ... fair play must be given to the machine. Instead of it sometimes giving no answer we could arrange that it gives occasional wrong answers. But the human mathematician would likewise make blunders when trying out new techniques. It is easy for us to regard these blunders as not

counting and give him another chance, but the machine would probably be allowed no mercy. In other words, then, *if a machine is expected to be infallible, it cannot also be intelligent.*[20]

Wrapped up in this passage are two of Turing's ideas about our kind of intelligence. One is the process of learning by trial and error, the way, say, that a baby learns to walk and talk. We make mistakes, but we learn from the mistakes and make fewer errors of the same kind as time passes. His dream was to have a computer prepared in a blank state, capable of learning about its environment and growing mentally as it did so. That dream is now becoming a reality, at least in a limited sense – for example, with robots that learn how their arms move by watching their own image in a mirror. The second idea concerns intuition, and the way humans can sometimes reach correct conclusions on the basis of limited information, without going through all the logical steps from A to Z. A computer programmed to take all those logical steps could never make the leap if some steps were missing.

Von Neumann shared this view of the importance of errors. In lectures he gave at Caltech in 1952, later published as a contribution to a volume edited by John McCarthy and Claude Shannon,[21] he said:

> Error is viewed, therefore, not as an extraneous and mis-directed or misdirecting accident, but as an essential part of the process.
>
> If the capacity to make mistakes of the kind just discussed is what distinguishes the human kind of intelligence from the machine kind of intelligence, would it ever be possible to program a classical computer, based on the principles

involved in the kind of machines discussed so far, to make deliberate mistakes and become intelligent like us? I think not, for reasons that will become clear in the rest of this book, but basically because I believe that the mistakes need to be more fundamental – part of the physics rather than part of the programming. But I also think it will indeed soon be possible to build non-classical machines with the kind of intelligence that we have, and the capacity for intellectual growth that Turing dreamed of.

Two questions that von Neumann himself raised are relevant to these ideas, and to the idea of spacefaring von Neumann machines:

Can the construction of automata by automata progress from simpler types to increasingly complicated types?

and

Assuming some suitable definition of efficiency, can this evolution go from less efficient to more efficient automata?

That provides plenty of food for thought about the future of computing and self-reproducing robots. But I'll leave the last word on Johnny von Neumann to Jacob Bronowski, no dullard himself, who described him as 'the cleverest man I ever knew, without exception … but not a modest man'. I guess he had little to be modest about.

First Interlude

Classical Limits

In the decades since EDSAC calculated the squares of the numbers from 0 to 99, computers have steadily got more powerful, faster and cheaper. Glowing valves have been replaced by transistors and then by chips, each of which contains the equivalent of many transistors; data storage on punched cards has been superseded by magnetic tape and discs, and then by solid state memory devices. Even so, the functioning of computers based on all of these innovations would be familiar to the pioneers of the 1940s, just as the functioning of a modern aeroplane would be familiar to the designers of the Hurricane and Spitfire. But the process cannot go on indefinitely; there are limits to how powerful, fast and cheap a 'classical' computer can be.

One way of getting a handle on these ideas is in terms of a phenomenon known as Moore's Law, after Gordon Moore, one of the founders of Intel, who pointed it out in 1964. It isn't really a 'law' so much as a trend. In its original form,

Moore's Law said that the number of transistors on a single silicon chip doubles every year; with another half-century of observation of the trend, today it is usually quoted as a doubling every eighteen months. And to put that in perspective, the number of transistors per chip has now passed the billion mark. That's like a billion-valve Manchester Baby or EDVAC on a single chip, occupying an area of a few hundred square millimetres.[1] At the same time, the cost of individual chips has plunged, they have become more reliable and their use of energy has become more efficient.

But there are problems at both ends of the scale. Although the cost of an individual chip is tiny, the cost of setting up a plant to manufacture chips is huge. The production process involves using lasers to etch patterns on silicon wafers on a tiny scale, in rooms which have to be kept scrupulously clean and free from any kind of contamination. Allowing for the cost of setting up such a plant, the cost of making a single example of a new type of chip is in the billions of dollars; but once you have made one, you can turn out identical chips at virtually no unit cost at all.

There's another large-scale problem that applies to the way we use computers today. Increasingly, data and even programs are stored in the Cloud. 'Data' in this sense includes your photos, books, favourite movies, emails and just about everything else you have 'on your computer'. And 'computer', as I have stressed, includes the Turing machine in your pocket. Many users of smartphones and tablets probably neither know nor care that this actually means that the data are stored on very large computers, far from where you or I are using our Turing machines. But those large computer installations have two problems. They need a lot of energy in the form of electricity; and because no machine is 100 per cent efficient they release a lot of waste energy, in the form of heat. So favoured locations for the physical machinery that represents the ephemeral image of the Cloud are places like Iceland and Norway, where there is cheap electricity (hydrothermal or just hydroelectric) and it is cold outside. Neither of these problems of the large scale is strictly relevant to the story I am telling here, but it is worth being aware that there must be limits to such growth, even if we cannot yet see where those limits are.

It is on the small scale that we can already see the limits to Moore's Law, at least as it applies to classical computers. Doubling at regular intervals – whether the interval is a year, eighteen months or some other time step – is an exponential process which cannot continue indefinitely. The classic example of runaway exponential growth is the legend of the man who invented chess. The story tells us that the game was invented in India during the sixth century by a man named Sissa ben Dahir al-Hindi to amuse his king, Sihram. The king was so pleased with the new game that he allowed Sissa to choose his own reward. Sissa asked for either 10,000 rupees or 1 grain of corn for the first square of the chess board, two for the second square, four for the third square and so on, doubling the number for each square. The king, thinking he was getting off lightly, chose the second option. But the number of grains Sissa had requested amounted to 18,446,744,073,709,551,615 – enough, Sissa told his king, to cover the whole surface of the Earth 'to the depth of the twentieth part of a cubit'. Alas, that's where the story ends, and we don't know what became of Sissa, or even if the story is true. But either way, the numbers are correct; and the point is that exponential growth cannot continue indefinitely or it would consume the entire resources not just of the Earth but of the Universe. So where are the limits to Moore's Law?

At the beginning of the twenty-first century, the switches that turned individual transistors on microchips on and off – the equivalent of the individual electromechanical relays in the Zuse machines or the individual valves in Colossus – involved the movement of a few hundred electrons. Ten years later, it involved a few dozen. We are rapidly[2] approaching the stage where an individual on–off switch in a computer, the binary 0 or 1 at the heart of computation and memory storage, is controlled by the behaviour of a single electron, sitting in (or on) a single atom; indeed, in 2012, while this book was in preparation, a team headed by Martin Fuechsle, of the University of New South Wales, announced that they had made a transistor from a single atom. This laboratory achievement is only a first step towards putting such devices on your smartphone, but it must herald a limit to Moore's Law as we have known it, simply because miniaturization can go no further; there is nothing smaller than an electron that could do the job in the same way. If there is to be future progress in the same direction, it will depend on something new, such as using photons to do the switching: computers based on optics rather than on electricity.

There is, though, another reason why the use of single-electron switches takes us beyond the realm of classical computing. Electrons are quintessentially quantum entities, obeying the rules of quantum mechanics rather than the rules of classical (Newtonian) mechanics. They sometimes behave like particles, but sometimes behave like waves, and they cannot be located at a definite point in space at a definite moment of time. And, crucially, there is a sense in which you cannot say whether such a switch is on or off – whether it is recording a 1 or a 0. At this level, errors are inevitable, although below a certain frequency of mistakes they might be tolerated. Even a 'classical' computer using single-electron switches would have to be constructed to take account of the quantum behaviour of electrons. But as we shall see, these very properties themselves suggest a way to go beyond the classical limits into something completely different, making quantum indeterminacy an asset rather than a liability.

In December 1959, Richard Feynman gave a now-famous talk with the title 'There's Plenty of Room at the Bottom',[3] in which he pointed the way towards what we now call nanotechnology, the ultimate forms of miniaturization of machinery. Towards the end of that talk, he said:

> When we get to the very, very small world – say circuits of seven atoms – we have a lot of new things that would happen that represent completely new opportunities for design. Atoms on a small scale behave like nothing on a large scale, for they satisfy the laws of quantum mechanics. So, as we go down and fiddle around with the atoms down there, we are working with different laws, and we can expect to do different things. We can manufacture in different ways. We can use, not just circuits, but some system involving the quantized energy levels, or the inter-actions of quantized spins, etc.

As I have mentioned, just half a century later we have indeed now got down to the level of 'circuits of seven atoms'; so it is time to look at the implications of those laws of quantum mechanics. And there is no better way to look at them than through the eyes, and work, of Feynman himself.

PART TWO

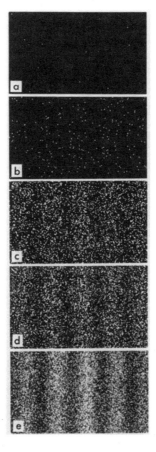

*The 'interference pattern' built up
by electrons passing one at a time
through 'the experiment with two holes'.
How do they know where to go?*

Chapter Three

Feynman and the Quantum

Richard Phillips Feynman was born on 11 May 1918, and grew up in Far Rockaway in the borough of Queens, New York. By the time he went to MIT, in 1935, the 'quantum revolution' of the 1920s was complete, and von Neumann had already written his influential book *The Mathematical Foundations of Quantum Mechanics*,[1] although at that time it had not been translated into English. To Feynman's generation, and later students, quantum mechanics was (and is) the received wisdom, not some startling new discovery, and that is the spirit in which I approach it here.

Feynman's father, Melville, had a fascination with science, especially natural history; he was intelligent and had wanted to become a doctor, but as the son of poor Lithuanian Jewish immigrants could not afford a college education. He ended up in the uniform business. Melville deliberately set out to encourage scientific interest in his son, buying him a set of the

Encyclopedia Britannica, taking him on trips to the American Museum of Natural History, and encouraging him to solve puzzles for himself rather than expecting to be given the answers. It turned out that Richard needed little encouragement, and had a natural aptitude for mathematics and, later, mathematical physics. The family were not affluent, but neither were they poor, surviving the Depression in relative comfort. At school, Richard was outstanding academically (at least in science and maths), often helping older students out with their assignments, but hopeless at ball games and self-conscious about his lack of what were perceived as 'manly' skills. He built radio receivers, repaired them for other people, learned to dance so that he could meet girls (he later said that as a teenager he was interested in only two things, maths and girls), and graduated from high school in the summer of 1935 with top honours.

Even so, the passage into college wasn't straightforward. Feynman applied to Columbia and MIT, but was rejected by Columbia because they still operated a quota on Jewish students, and had filled this already. MIT had a different hoop that had to be jumped through – applicants had to have a recommendation from an MIT graduate before they would be considered. Melville persuaded an acquaintance to provide the endorsement; Feynman later described this system as 'evil, wrong, and dishonest'.[2]

MIT

Feynman's reputation as a budding scientist preceded him to MIT, where he was the subject of rivalry between the only two Jewish fraternities, Phi Beta Delta and Sigma Alpha Mu, each eager to add him to its number. Although Feynman had no religious beliefs, his family background meant that he had to join one or other of these fraternities; so he settled on Phi Beta Delta, partly because two older members had advised him that as an outstanding student he would be allowed to take examinations on arrival at MIT which, once passed, would enable him to skip the first-year maths lectures and start with the second-year course. As this shows, the frats were not all about partying, but provided a mutual support system for members. For example, the more academic students were expected to help the more social animals with their work, and in return the social types helped the academics to come out of their shells and learn the social graces. Feynman described it as 'a good balancing act' from which he benefited, losing the self-consciousness that had handicapped him in high school.

Feynman benefited in another way from living in the frat house. Two of the senior students there[3] were taking an advanced course in physics which included the latest developments in quantum mechanics. Through conversations with them, Feynman decided to switch from mathematics to physics, and signed up for the same physics course (intended for seniors and graduate students) at the start of his second (sophomore) year. Even in this advanced company, Feynman stood out. For the first semester, the course was taught by a young professor, Julius Stratton, who later became President of MIT, but in 1936 was sometimes careless about preparing his lectures. Whenever he got stuck, he would turn to the audience and ask, 'Mr Feynman, how did you handle this problem?' and Richard would take over.[4] Nobody else in the class was ever singled out in this way.

Along with advanced physics, during his time as an

undergraduate Feynman took regular courses in chemistry, metallurgy, experimental physics and optics, and signed up for another advanced course in nuclear physics. He sailed through anything scientific. But he only passed the compulsory courses in English, history and philosophy, which he regarded as 'dippy' subjects, with the aid of the fraternity support system. He published two scientific papers before he had even graduated, and wanted to stay on at MIT to work for a PhD, but was told it would be better for his scientific development to go somewhere else. Somewhat grudgingly, he complied, moving in 1939 from MIT to Princeton; he later acknowledged that his teachers were right, and that the move was the right thing for him at that time.

From Princeton to Los Alamos

Princeton had been alerted that something special was coming their way, but even so nearly turned him down when they saw his grades. He had scored 100 per cent in physics, and almost as high in maths. Both were the best scores the Princeton Graduate Admissions Committee had ever seen. But they had never admitted anybody with such low scores as Feynman had achieved (if that is the right word) in English and history. In the end, he was offered a research assistantship, which meant that he worked for a senior scientist and actually got paid while doing his own PhD research. The scientist Feynman worked with was John Wheeler, later famous for his investigations of the physics of black holes. But 'senior' is a relative term: when they met, Wheeler was twenty-eight and Feynman twenty-one. They became good friends, and Wheeler also acted as Feynman's thesis adviser.

Feynman's thesis was entitled 'The Principle of Least

Action in Quantum Mechanics', and dealt with a way of describing how quantum entities such as electrons travel from A to B. This led to the so-called 'path integral approach', and to the work for which Feynman would later receive the Nobel Prize. I shall explain all this shortly; but Feynman's career was interrupted, just at the point he was finishing his thesis in 1941, by the involvement of the United States in the Second World War.

Even before the attack on Pearl Harbor, like many of his contemporaries Feynman had realized that war was inevitable, and in the summer of 1941 had been working at the Frankfort Arsenal in Philadelphia on a mechanical predictor for anti-aircraft gunnery. He made such an impression that he was offered a full-time job at the head of his own design team, but went back to Princeton to finish his PhD; had his decision gone the other way, he might well have become a leading light in the early development of electronic computers. Feynman was initially recruited to war work in December 1941, at first on the problem of separating radioactive uranium-235 from the stable variety, uranium-238. This was before he had completed his thesis, but he took a few weeks' leave in the spring of 1942 to write it up. The oral examination, held on 3 June 1942, was a formality, and he received the degree the same month. Before June was out, Richard also married his childhood sweetheart, Arline Greenbaum, even though she was seriously ill (indeed, hospitalized) with tuberculosis. Later that year, the uranium enrichment project that Feynman was involved with was dropped, in favour of a more successful method, and he was moved, along with other members of the team, to Los Alamos, where, among other things, he worked with the IBM machines needed to help von Neumann with his calculations. Arline also moved west, to a hospital as close as possible to Los Alamos, where she died in 1945.

Before picking up the threads of Feynman's work on quantum physics after the war, and in particular his prescient ideas about computing and quanta, we can put all this in context by looking at the kind of quantum physics he had been taught as an undergraduate – the kind of quantum physics espoused by von Neumann in his book. In its most widely used form, this was based on an equation discovered by the Austrian Erwin Schrödinger.

Schrödinger and his equation

One of the peculiarities of quantum physics is that although we have very good, reliable equations to describe what is going on in the subatomic world, we do not have a single clear understanding of what it is those equations describe. The problem is not that we have *no* picture of what is going on, but that we have *many*, equally valid, pictures. There are several different ways of interpreting the equations in terms of the behaviour of quantum entities such as electrons, all of them equally valid in the sense that they are described by equations which allow physicists to make accurate and correct predictions about the outcome of experiments. I have gone into the details of all this in my book *Schrödinger's Kittens*; here, I shall mention just one (but the most profound) aspect of this intriguing puzzle.

In the middle of the 1920s, two completely different ways of understanding the quantum world were developed in-dependently, at almost the same time. The first approach, stemming from the work of Werner Heisenberg, treated electrons as particles, whose behaviour could be described with great precision by a certain set of equations and mathematical rules. At one level, this matches the idea most of us have of electrons as tiny, subatomic particles, like little billiard balls, each carrying a certain amount of electric charge. To be sure, there were some oddities about the rules, not least that the 'particles' could jump from one place to another instantaneously, without crossing the space in between. But the equations worked. The second approach, developed initially by Schrödinger, treated electrons as waves. This meant that they could be described in terms of the rules of wave behaviour, which physicists were confident they knew all about from studying things like ripples in water. To be sure, there were some oddities about the picture, not least the puzzle of how the electric charge of the electron could be carried by a wave. But the equations worked.

Very soon, several people (most notably Paul Dirac) proved that both these versions of quantum mechanics (and, indeed, all versions of quantum mechanics) are mathematically equivalent to one another, not unlike the way in which a book like this might be written in English and also in German or some other language but still contain, and convey, the same message. This meant that in a sense it didn't matter which version you chose to work with, since they all gave the same answers. Because physicists were already familiar with the idea of waves and wave equations, Schrödinger's version of quantum mechanics quickly became the most popular, and was developed into a standard version which became known as the Copenhagen Interpretation, because one of the leading proponents of the idea, Niels Bohr, worked in Copenhagen. This is the version that I am going to tell you about now, the version Richard Feynman learned as a student; but you should not imagine for one moment that it is the ultimate truth about quantum physics, or that electrons 'really are' waves. If you only want to do calculations about the outcome of experiments involving subatomically tiny things like electrons it works fine; and for half a century hardly anybody worried about what the quantum world is 'really' like. In a memorable phrase coined by John Bell (of whom more shortly), the quantum world behaves 'for all practical purposes' (FAPP) as if electrons were waves obeying the Schrödinger equation as interpreted by the Copenhagen school.

The fundamental feature of the Copenhagen Interpretation is that a quantum entity such as an electron can be represented by a wave, described by a wave equation (also known as a 'wave function'). This wave occupies a large volume of space (potentially, an infinitely large volume). The wave function has a value at any point in space, and this number is interpreted, following a suggestion made by Max Born, as representing the probability of finding the electron at that point. In some places, the wave is, in a sense, strong (the number associated with the wave function is large), and there is a high probability that if we look for the electron we will find it in one of those places; in other places the wave is weak, and there is a small probability of finding the electron in one of those places. But when we look for the electron we do find it in a definite place, like a particle, not as a spread-out wave. The wave function is said to 'collapse' onto that place. But as soon as the experiment is over, the wave starts spreading out across the Universe. It is this combination of waves, probability and collapse which makes up the Copenhagen Interpretation, and which von Neumann wrapped up in one neat package in his book.

It's worth spelling this out in detail. Imagine that we have a single electron confined in a large box. According to the Copenhagen Interpretation, the wave function fills the box evenly – the chance of finding the electron at any point in the box is the same as the chance of finding it anywhere else in the box. Now, we make a measurement to detect the electron. We find it, looking just like a little particle, at a definite point in the box.[5] But as soon as we stop monitoring the electron, the wave function immediately spreads out from the point where we discovered it. If we quickly make another measurement, there is a high probability of finding the electron close to the place where we last saw it. That matches common sense – but there is still some quantifiable probability of finding it anywhere in the box. The longer we wait, the more the wave function develops, and the chances of finding the electron *anywhere* in the box even out. That's weird, but not completely crazy.

But it is just the beginning. Richard Feynman was fond of presenting what he called 'the central mystery' of quantum mechanics by applying the Copenhagen Interpretation to a description of what happens to an electron (or any other quantum entity) when it passes through what he called 'the experiment with two holes'. It 'has in it', he said, 'the heart of quantum mechanics'.[6]

The experiment with two holes

What Feynman called 'the experiment with two holes' is more formally known as the double-slit experiment, and may be familiar from your schooldays, for it is often used to demonstrate the wave nature of light. In this version, light is shone through a small hole in a darkened room, and spreads out from the hole to fall upon a screen (maybe just a sheet of cardboard) in which there are two holes, ether pinpricks or parallel razor slits. Beyond this screen there is another sheet of cardboard, where the light spreading from the two holes makes a pattern. For the double-slit version of the experiment, the pattern is one of parallel dark and bright stripes, and is explained in terms of the interference of waves spreading out from each of the two slits. You can see exactly the same kind of wave interference if you drop two stones into a still pond simultaneously, although that pattern corresponds to the 'double-pinprick' version of the experiment. All this is compelling evidence that light travels as a wave. But there is also compelling evidence that in some circumstances light behaves like a stream of particles. This is the work for which Albert Einstein received his Nobel Prize. One way of observing this is to replace the final sheet of cardboard in the experiment with a screen like that in a TV, and turn the brightness of the beam of light down really low. Now, individual 'particles of light' (photons) can be seen arriving at the screen, where they make little flashes, each at a definite (within the limits of quantum uncertainty) point. It looks as if single particles are arriving at the detector, one at a time. But if you record the process over a long period of time, it becomes clear that the flashes of light are occurring more frequently in some parts of the screen than others; indeed, they build up into the usual interference pattern of light and dark stripes. Somehow, the 'particles' are conspiring to produce the pattern we expect for waves.

It gets weirder. If the photons 'really are' particles, they ought to produce a quite different pattern. Imagine firing a stream of bullets through the two slits (in an armour-plated screen!) into a sandbank. There would be one pile of spent bullets behind each slit, and nothing anywhere else. So which pattern would you expect if you fired a beam of electrons through an equivalent experiment? A team from the Hitachi research laboratories and Gakushin University in Tokyo did just that in 1987. The results were exactly the same as the results for photons. The beam of electrons interfered with itself to produce the familiar pattern corresponding to waves. And when the power of the beam was turned down so low that electrons were leaving the emitter one at a time, they produced individual flashes on the screen, which built up to make the interference pattern. Essentially, the same quantum weirdness applies to electrons and to light.

It doesn't end there. It is also possible to set up the double-slit experiment for electrons in such a way that we can tell which of the two slits an individual electron goes through. When we do this, we do not get an interference pattern on the final screen; we get two blobs of light, one behind each slit, equivalent to the heaps of spent bullets. The electrons seem to be aware that they are being monitored, and adjust their behaviour accordingly. In terms of the collapse of the wave function, you can say that what happens is that by looking at the hole we make the wave function collapse into (or onto) a particle, affecting its behaviour. This almost makes sense. Curiously, though, we only have to look at one of the two slits for the outcome of the whole experiment to be affected, as if the electrons passing through the other slit also knew what we were doing. This is an example of quantum 'non-locality', which means that what happens in one location seems to affect events in another location instantly.

Non-locality is a key feature of the central mystery of quantum mechanics, and a vital ingredient in quantum computers.

Well, you may say, nobody has ever seen an electron, and we can't really be sure that we have interpreted what is going on correctly. But in 2012 a large team of researchers working at the University of Vienna and the Vienna Centre for Quantum Science and Technology reported that they had observed the same kind of matter–wave interference using molecules of a dye, phthalocyanine, that are so large (0.1 mm across) that they can be seen with a video camera. Just as in the case of light, electrons, and also individual atoms used in other studies, the interference pattern characteristic of waves builds up even when the molecules are sent one at a time through the experiment with two holes. And it disappears if you look to see which hole the particles go through. The 'central mystery' of quantum mechanics writ large, literally.

Feynman came up with a way to explain what is going on, and to extend it into a broader understanding of quantum reality, in what became his PhD thesis.

Integrating history

One way of interpreting what is going on, valid 'FAPP', is to calculate the behaviour of a wave of probability passing through the experiment with two holes and interfering with itself to determine where particles are allowed to arrive on the final screen. This is a straightforward thing to do using wave mechanics, and leads to the standard pattern of bright and dark stripes. In some places, the probabilities reinforce each other, and these correspond to places where the electron might be found; in other places, the probabilities cancel out, and there is no chance of finding the electron there. If you imagine cutting four parallel, equally spaced slits in the middle screen instead of two, you can do the equivalent, slightly more complicated calculation and work out the corresponding pattern. With eight slits, we would have to add up eight lots of probabilities to determine the pattern, and so on.

Perhaps you can see where we are going. Even with a million razor slits, you could, in principle, calculate the pattern of bright and dark places on the final screen. That would correspond, FAPP, to each single electron going through a million holes at once. Why stop there, asked Feynman? Why not take the screen away altogether, leaving an infinite number of paths for the electron to follow from one side of the experiment to the other? It is actually easier to calculate the result for such a situation than for one with a million slits, because the mathematical rules make it straight-forward to work out the implications in the limit; that is, where the numbers approach infinity. Without actually doing an infinite number of calculations, it is possible to work out which kinds of paths combine together and which ones cancel each other out. The probabilities for more complicated paths turn out to be very small, and also to cancel each other out. Only a small number of possible paths, very close to one another, reinforce each other; they combine to produce a single spot on the final detector screen. The interference pattern disappears, and we are left with what looks like a classical particle travelling from one side of the experiment to the other along a single path, or trajectory. The process of adding up the probabilities for each path is known as the 'path integral' approach, or sometimes as the 'sum over histories' approach.

This sounds like a mere mathematical trick. But it is actually possible to see light travelling by some of these 'non-classical' paths. All you need is a compact disc. One of the other things we learned in school is that light travels in straight lines, so that when it encounters a mirror it bounces off at the same angle that it arrives – the angle of reflection is equal to the angle of incidence. But that isn't the whole story. According to Feynman's path integral approach, when light bounces off a mirror it does so at all possible angles, including crazy paths where it arrives perpendicular to the mirror and reflects at a shallow angle to meet your eye, and paths where it arrives at a grazing angle and bounces off at a right angle to meet your eye. All the 'crazy' paths cancel each other out; only paths near to the shortest distance from the light to the mirror to your eye reinforce each other, leaving the appearance of light travelling in a single straight line. But the 'crazy' paths really are there. They cancel out because, except near the classical path, light waves (or probability waves) in neighbouring strips of the mirror are out of step with one another (out of phase). In one strip, the probabilities go one way, but in the next strip they go the other way. If we carefully lay strips of black cloth over regions where the probabilities all point one way, we are left with parallel strips of mirror, separated by the strips of cloth, where the probabilities all point the other way, so there is no cancelling. The spacing of the strips needed to make this work depends on the wavelength of the light involved, so it is related to the colour of the light (red light has a longer wavelength than blue light).

It really is possible to set up a simple experiment with a light source, a mirror and an observer (your eye!), so that when you choose a part of the mirror where there is no visible reflection, then cover up strips of this part of the mirror in the right way, you will see a reflection. With part of the mirror covered up, it really does seem as if less is more when it comes to seeing reflections.

But you don't need to go to the trouble of doing this experiment to see crazy reflections. The grooves in a CD are like little strips of mirror separated by regions where there is no reflecting material, and it happens that the spacing of these grooves is just right for the effect to work. If you hold a CD under a light, you don't just see a simple image of the light as you would from an ordinary mirror; you also see a coloured, rainbow pattern of light from right across the disc. The rainbow pattern is because different wavelengths are affected slightly differently, but you can see reflected light coming from 'impossible' regions of the disc, just as the path integral approach predicts. But even with a conventional mirror, 'light doesn't *really* travel only in a straight line,' says Feynman, 'it "smells" the neighbouring paths around it, and uses a small core of nearby space'.[7]

What is special about that 'small core' of space? Why does light 'move in straight lines'? It's all to do with something called the Principle of Least Action, which also intrigued Feynman. Indeed, it was the Principle of Least Action that started him on the path which led to his Nobel Prize.

A PhD with a principle

Feynman had actually learned about this principle when he was still in high school, from a teacher, Abram Bader, who appreciated his unusual ability and encouraged him to go beyond the regular syllabus. It can best be understood in terms of the flight of a ball thrown from ground level through an open window on the upper floor of a house. At any point along its trajectory, the ball possesses both kinetic energy, due to its motion, and gravitational potential energy, related to its height above the ground. The sum of these two energies is always the same, so the higher the ball goes the more slowly it moves, trading speed for height. But the *difference* between the two energies changes as the ball moves along its path. 'Action', in the scientific sense, relates these changing energies to the time it takes for the ball to complete its journey. The difference between the kinetic and potential energies can be calculated for any point of the trajectory, and the action is the sum of all these differences, integrated along the whole trajectory. Equivalent actions can be calculated for other cases, such as a charged particle moving under the influence of an electric force.

The fascinating fact which the intrigued Feynman learned from Mr Bader is that the trajectory followed by the ball (which, you may recall, is part of a parabola), is the path for which the action is least. And the same is true in general for other cases, including for electrons moving under the influence of magnetic or electric forces. The trajectory corresponding to least action is also the one corresponding to least time – for any starting speed of the thrown ball, the appropriate parabola describes the path for which the ball takes least time to get to the window. Anyone with experience of throwing balls knows that the faster you throw the flatter the trajectory has to be to hit such a target, and this is all included in the Principle of Least Action. In the guise of the Principle of Least Time, it can also be applied to light. Light always travels in straight lines, we are taught, so we don't think of it as following a parabola from the ground through an upper-storey window. But it changes direction when it encounters a different material, such as when it moves from air into a glass block. The light travels in a straight line through the air up to the edge of the glass, then changes direction and travels in another straight line through the glass. The path it follows, getting from point A outside the glass to point B inside the glass, is always the one that takes least time. Which is *not* the shortest distance, because light travels more quickly in air than it does in glass. Just as with the ball thrown through the window, it is the whole path that is involved in determining the trajectory.

All this applies to 'classical' (that is, non-quantum) physics. Feynman's contribution was to incorporate the idea of least action into quantum physics, coming up with a new formulation of quantum physics different from, and in many ways superior to, those of the pioneers, Heisenberg and Schrödinger. This was something which might have been the crowning achievement of a lesser scientist, but in Feynman's case was 'merely' his contribution as a PhD student, completed in a rush before going off to war work at Los Alamos.

In his Nobel Lecture,[8] Feynman said that the seed of the idea which became his PhD thesis was planted when he was an undergraduate at MIT. At that time, the problem of an electron's 'self-interaction' was puzzling physicists. The strength of an electric force is proportional to 1 divided by the distance from an electric charge, but the distance of an electron from itself is 0, and since 1 divided by 0 is infinity, the force of its self-interaction ought to be infinite. 'Well,' Feynman told his audience in Stockholm, it seemed to me quite evident that the idea that a particle acts on itself, that the electrical force acts on the same particle that generates it, is not a necessary one – it is a sort of a silly one, as a matter of fact. And so, I suggested to myself that electrons cannot act on themselves; they can only act on other electrons . . . It was just that when you shook one charge another would shake later. There was a direct interaction between charges, albeit with a delay . . . Shake this one, that one shakes later. The sun atom shakes; my eye electron shakes eight minutes later, because of a direct interaction.

The snag with this idea was that it was too good. It meant that when an electron (or other charged particle) interacted with another charged particle by ejecting a photon (which is the way charged particles interact with one another) there would be no back-reaction to produce a recoil of the first electron. This would violate the law of conservation of energy, so there had to be some way to provide just the right amount of interaction to produce a kick of the first electron (equivalent to the kick of a rifle when it is fired[9]) without being plagued by infinities. Stuck, but convinced that there must be a way round the problem, Feynman carried the idea with him to Princeton, where he discussed it with Wheeler; together they came up with an ingenious solution.

The starting point was the set of equations, discovered by James Clerk Maxwell in the nineteenth century,[10] which describe the behaviour of light and other forms of electro-magnetic radiation. It is a curious feature of these equations that they have two sets of solutions, one corresponding to an influence moving forward in time (the 'retarded solution') and another corresponding to an influence moving backward in time (the 'advanced solution'). If you like, you can think of these as waves moving either forward or backward in time, but it is better to avoid such images if you can. Since Maxwell's time, most people had usually ignored the advanced solution, although mathematically inclined physicists were aware that a combination of advanced and retarded solutions could also be used in solving problems involving electricity, magnetism and light. Wheeler suggested that Feynman might try to find such a combination that would produce the precise feedback he needed to balance the energy budget of an emitting electron.

Feynman found that there is indeed such a solution, provided that the Universe absorbs all the radiation that escapes out into it, and that the solution is disarmingly simple. It is just a mixture of one-half retarded and one-half advanced interaction. For a pair of interacting electrons (or other charged particles), half the interaction travels forward in time from electron A to electron B, and half travels backward in time from electron B to electron A. As Feynman put it, 'one is to use the solution of Maxwell's equation which is sym-metrical in time'. The overall effect is to produce an interaction including exactly the amount of back-reaction (the 'kick') needed to conserve energy. But the crucial point is that the whole interaction has to be considered as – well, as a whole. The entire process, from start to finish, is a seamless and in some sense timeless entity. This is like the way the whole path of the ball thrown through the window has to be considered to determine the action, and (somehow!) for nature to select the path with least action. Indeed, Feynman was able to reformulate the whole story of interacting electrons in terms of the Principle of Least Action.

The discovery led to Feynman's thesis project, in which he used the Principle of Least Action to develop a new understanding of quantum physics. He started from the basis that 'fundamental (microscopic) phenomena in nature are symmetrical with respect to the interchange of past and future' and pointed out that, according to the ideas I have just outlined, 'an atom alone in empty space would, in fact, not radiate . . . all of the apparent quantum properties of light and the existence of photons may be nothing more than the result of matter interacting with matter directly, and according to quantum mechanical laws', before presenting those laws in all their glory. The thesis, he emphasized, 'is concerned with the problem of finding a quantum mechanical description applicable to systems in which their classical analogues are expressible by a principle of least action'. He was helped in this project when a colleague, Herbert Jehle, showed him a paper written by Paul Dirac[11] that used a mathematical function known as the Lagrangian, which is related mathematically to action, in which Dirac said that a key equation 'contains the quantum analogue of the action principle'. Feynman, being Feynman, worried about the meaning of the term 'analogue' and tried making the function equal to the action. With a minor adjustment (he had to put in a constant of proportionality), he found that this led to the Schrödinger equation. The key feature of this formulation of quantum mechanics, which many people (including myself) regard as the most profound version, is, as Feynman put it in his thesis, that 'a probability amplitude is associated with an entire motion of a particle as a function of time, rather than simply with a position of a particle at a particular time'. But everything, so far, ignored the complications caused by including the effects of the special theory of relativity.

This is more or less where things stood when Feynman wrote up his thesis and went off to Los Alamos. After the war, he was able to pick up the threads and develop his ideas into a complete theory of quantum electrodynamics (QED) including relativistic effects, the work for which he received a share of the Nobel Prize; but this is not the place to tell that story.

What I hope you will take away from all this is the idea that at the quantum level the world is in a sense timeless, and that in describing interactions between quantum entities everything has to be considered at once, not sequentially. It is not necessary to say that an electron shakes and sends a wave moving out across space to shake another electron; it is equally valid to say that the first electron shakes and the second one shakes a certain time later as a result of a direct, but delayed, interaction. And the trajectory 'sniffed out' when an electron moves from A to B is the one that corresponds to the least action for the journey. Nature is lazy.

Nobody said this more clearly than Feynman, as another quote from his Nobel Lecture shows:

> We have a thing [the action] that describes the character of the path through all of space and time. The behavior of nature is determined by saying her whole space–time path has a certain character . . . if you wish to use as variables only the coordinates of particles, then you can talk about the property of the paths – but the path of one particle at a given time is affected by the path of another at a different time.

This is the essence of the path integral formulation of quantum mechanics.

Intriguingly, Schrödinger's quantum wave function has built into it exactly the same time symmetry as Maxwell's equations, which Schrödinger himself had commented on in 1931, but had been unable to interpret, simply remarking: 'I cannot foresee whether [this] will prove useful for the explanation of quantum mechanical concepts.' Feynman doesn't seem to have been influenced by this, and may not have been aware, in 1942, of Schrödinger's remark, which was made in a paper published in German by the Prussian Academy of Science. Schrödinger was also ahead of Feynman, although in an almost equally understated way, in appreciating that all quantum 'paths' are equally valid (but not necessarily equally probable). A convenient way into his version of this idea is through Schrödinger's famous cat puzzle (sometimes called a 'paradox', although it isn't one).

Cats don't collapse

The point of Schrödinger's 'thought experiment', published in 1935, was to demonstrate the absurdity of the Copenhagen Interpretation, and in particular the idea of the collapse of the wave function. His hypothetical cat[12] was imagined to be living happily in a sealed chamber, on its own with plenty to eat and drink, but accompanied by what Schrödinger called a 'diabolical device'. This device could involve any one of several quantum systems, but the example Schrödinger chose was a combination of a sample of radioactive material and a detector. If the sample decays, spitting out an energetic particle, the detector triggers a piece of machinery which releases poison and kills the cat. The point is that the device can be set up in such a way that after a certain interval of time there is a 50:50 chance that the radioactive material has or has not decayed. If the sample has decayed, the cat dies; if not, the cat lives. But the Copenhagen Interpretation says that the choice is not made until someone observes what is going on. Until then, the radioactive sample exists in a 'super-position of states', a mixture of the two possible wave functions. Only when it is observed does this mixture collapse one way or the other. This is equivalent to the way the wave associated with an electron going through the experiment with two holes goes both ways at once through the slits when we are not looking, but collapses onto one or the other of the slits when we put a detector in place to monitor its behaviour.

This is all very well for electrons, and even radioactive atoms. But if we take the Copenhagen Interpretation literally it means that the cat is also in a mixture of states, a super-position of dead and alive, and collapses into one or the other state only when someone looks to see what is going on inside the room. But we never see a cat that is dead and alive (or neither dead nor alive) at the same time. This is the so-called paradox.

It is possible to extend this idea beyond Schrödinger's original formulation to bring out another aspect of quantum reality that is crucial in understanding quantum computation. Instead of one cat, imagine two (perhaps brother and sister), each in its own chamber, both connected to a diabolical device which determines, with 50:50 precision, that as a result of the radioactive decay one cat dies and the other lives. Now, after the mixture of states has been set up, the two chambers, still sealed, are separated and taken far apart from each other (in principle, to opposite sides of the Galaxy). The Copenhagen Interpretation says that *each* chamber contains a superposition of dead and alive cat states, until someone looks in *either one* of the chambers. As soon as that happens, the wave function collapses for *both* cats, instantaneously. If you look in just one chamber and see a dead cat, it means that the wave function for the other cat has collapsed at that precise instant to form a live cat, and vice versa. But the quantum rules do *not* say that the collapse happened before the chambers were separated and that there 'always was' a dead cat in one chamber and a live cat in the other. Like the way a measurement at one slit in the experiment with two holes seems to affect what is going on at the other slit, this is an example of what is known as 'quantum non-locality', of which more shortly. But Schrödinger's resolution of all these puzzles was to say that there is no such thing as the collapse of the wave function. As early as 1927, at a major scientific meeting known as a Solvay Congress, he said: 'The real system is a composite of the classical system in all its possible states.' At the time, this remark was largely ignored, and the Copenhagen Interpretation, which worked For All Practical Purposes, even if it doesn't make sense, held sway for the next half-century. I will explain the importance of his alternative view of quantum reality later; but Schrödinger was way ahead of his time, and it is worth mentioning now that in 2012, eighty-five years after he made that remark, two separate teams offered evidence that wave functions are indeed real states that do not collapse. Before coming up to date, though, we should take stock of what Feynman and his contemporaries had to say about quantum computation.

The gateway to quantum computation

In order to understand their contribution, we need just a passing acquaintance with the logic of computation. This is based on the idea of logic 'gates': components of computers which receive strings of 1s and 0s and modify them in accordance with certain rules. These are the rules of so-called Boolean logic (or Boolean algebra), developed by the mathematician George Boole in the 1840s. Boolean algebra can be applied to any two-valued system, and is familiar to logicians in the form of application to true/false systems; but in our context it applies to the familiar binary language of computation. When we talk blithely about computers carrying out the instructions coded in their programs, we are really talking about logic gates taking 1s and 0s and manipulating them in accordance with the rules of Boolean algebra.

These rules are very simple, but not quite the same as those used in everyday arithmetic. For example, in everyday arithmetic 1 + 0 is always equal to 1. But using Boolean algebra a so-called AND gate will take an input of 1 + 0 and give the output 0. It will do the same for 0 + 0 and for 0 + 1 (which is different from 1 + 0 in the Boolean world). It will only give the output 1 if *both* inputs are 1; in other words if input A *and* input B are both 1, which is where it gets its name. Another gate, called the NOT gate, will always give the opposite output from its input. Put in 1 and out comes 0; put in 0 and out comes 1. The output is *not* the same as the input. Don't worry about how the possible kinds of gates combine to carry out the instructions in a computer program; I just want you to be aware that such gates exist and give you the simplest idea of what is actually going on inside your smartphone or other computer. An important part of developing computers is to devise gates which do interesting things in a reliable way, and in the mid-1970s this endeavour led to a major rethinking of the possibilities of machine computation.

One of the questions that computer scientists working at the more esoteric end of their speciality asked in the 1950s and 1960s was whether it was possible, in principle, to build a computer that could simulate precisely (not just approximately; no matter how good an approximation is, it is never perfect) the workings of 'classical' physics – the physics of colliding billiard balls and orbital motion of planets described so beautifully by Newton's laws. This brought them up against one of the most intriguing features of the world. Newton's laws are reversible. If we ignore things like friction, a collision between two individual pool balls, for example, looks the same going forwards in time or backwards in time. If we made a movie of the collision, and cut out the player making the shot, it would make as much sense backwards as it does forwards. But although each individual collision is reversible, if we made a similar movie showing the break in a game of pool, it would be obvious which was the future and which the past. The future is the situation where there is more disorder. This is a simple example of one of the most fundamental laws in science, the second law of thermodynamics. This says that, left to its own devices, the amount of disorder in a system (measured by a quantity called entropy) always increases, in spite of the reversibility of Newton's laws. The relevance of these ideas to computation comes about because entropy can be measured in terms of information. Another very simple example helps. If we have a glass containing water and some ice cubes, it takes more information to describe than if we have a glass that contains just water, even if it is the same glass but the ice has now melted – and once again, the direction of the 'arrow of time' is clear. The greater the entropy of a system, the more disordered it is, and the less information it contains.

The relationship between information (more specifically, information *transmission*) and thermodynamics was put on a mathematical basis by Claude Shannon, working at the Bell Telephone Laboratories in the 1940s. As this affiliation suggests, the idea was developed in the context of telephone communications, rather than computation, but it was soon taken up by the computer scientists. In this context, information is always a measure of the decrease of uncertainty provided by a message or a computation, and the direct motivation for Shannon's work was to find out how many calls could be sent simultaneously down a telephone cable without unacceptable loss of information. Incidentally, Shannon, who had been born in 1916, made his name developing telephone switching relay systems; during the Second World War he had worked at Bell Labs on cryptography and gunnery control systems, and met Alan Turing when he visited the labs in 1943. Shannon moved to MIT in 1956 and spent the rest of his career there.

Now, it is important to appreciate in what follows that we are talking about only those entropy changes involved in the act of computation itself, not the ones involved in the process of producing electricity for the computer to run on, or air conditioning for the room in which it operates. In doing so we are following the same convention physicists use in discussing the movements of frictionless pool balls to get insight into Newton's laws, and it leads to equally deep truths. But at first computer scientists were led in the wrong direction – ironically, by Johnny von Neumann, who gave a lecture in 1949 in which he calculated that there must be a minimum amount of energy required in the basic act of computation, that is, switching a 0 into a 1 or vice versa. And this would make it impossible to simulate the world of classical physics precisely, because not only would it take energy (and increase entropy) to run a calculation in a computer, energy would be required and entropy would increase still more if the computer were run backwards. The operation would not be reversible – at least, not in entropy/information terms.

The first hint that this was not the whole story came in 1961, when IBM researcher Rolf Landauer pointed out that at least some aspects of computation need not involve dissipating energy at all. Landauer was one of the first people to appreciate that, as he put it, 'information is physical', and that the unspoken assumption that there was some abstract, 'pure' form of computation that exists regardless of the machinery used is wrong. As he later said:

> Information is inevitably tied to a physical representation. It can be engraved on stone tablets, denoted by a spin up or spin down, a charge present or absent, a hole punched in a card, or many other physical phenomena. It is not just an abstract entity; it does not exist except through physical embodiment. It is, therefore, tied to the laws of physics.[13]

It is in that spirit that we can understand Landauer's insight of 1961. Imagine that the state of a single bit of information is represented by the presence of a ball in one of two holes, of equal depth, separated by a small hill. If the ball is in hole A, the bit is 1; if it is in hole B, the bit is 0. To change the state of the bit, the ball has to be rolled up the hill and down the other side. But the energy needed to raise the ball up one side of the hill is exactly equal to the amount of energy released when it is lowered down the other side. So, in principle, a computation can be carried out without using any energy at all! Another way of visualizing a change which does not use up energy is to think of a skateboarder at the top of a (frictionless) half-pipe. Starting out with zero speed, the skateboarder accelerates down to the bottom of the pipe then decelerates up the other side, arriving at the top with zero speed. He or she has changed position without using any energy.

This, though, is not the end of the story. Imagine that we started with a whole array of balls each in its equivalent of hole A, like a computer starting out from a state with every bit registering 1, then performed a computation (using zero energy) which left it with some balls in 'A' holes and some in 'B' holes. Landauer also showed that reversing the process of computation to erase the information (in computer jargon, resetting the register to its initial state) sometimes *does* require energy. This is a little more tricky to understand, but depends on the fact that the computer has to operate some process which will restore it to an original situation regardless of the present state. In the example I have used, the original starting situation involves all the balls being in their starting positions. Clearly, the ball in hole B can be put back into hole A by using energy which pushes it back over the hill – energy which can again be reclaimed as it falls down the other side. But the computer does not 'know' whether the ball is in hole B, so even if the ball is not there, the energy needed to reset the bit has to be applied, just in case it was there. Sometimes, this energy is wasted. This is the key insight provided by Landauer: that the computation itself need not involve any dissipation of energy (and so is, in principle, reversible), but that, paradoxical though it may seem, energy is dissipated every time that information is discarded!

Another way of looking at this, in terms of reversibility of information rather than dissipation of energy, is that if there is more than one way in which a state can be reached; that is, if it can be reached by different routes, the computer does not 'know' which is the right path to follow in order to reset the register – and this brings us back to the logic of gates. The AND gate is a good example. If we are trying to reset the registers and come across an AND gate in the state 1, we know for sure that the original input to the gate was 1 + 1. But if we find the AND gate in the state 0, we have no idea whether the input was 0 + 0, 1 + 0, or 0 + 1. A reversible computer, one which can simulate classical physics perfectly, has to be built entirely out of reversible gates. And such gates do exist.

Fredkin, Feynman and friends

The next step in the theory of reversible computation came from Charles Bennett, another IBM researcher, in 1973. Inspired by reading Landauer's paper and hearing him talk a few years earlier, he wrote some very simple computer programs which were indeed reversible – that is, in carrying out the exercise he realized that in every case the computation consisted of two halves, the second half almost exactly undoing the work of the first half. As he later explained:

> The first half would generate the desired answer . . . as well as, typically, some other information . . . The second half would dispose of the extraneous information by reversing the process that generated it, but would keep the desired

answer. This led me to realize that any computation could be rendered into this reversible format by accumulating a history of all information that would normally be thrown away, then disposing of this history by the reverse of the process that generated it. To prevent the reverse stage from destroying the desired output along with the undesired history, it suffices, before beginning the reverse stage, to copy the output on blank tape. [This] copying onto blank tape is already logically reversible.[14]

This is as if you had a tablet (in his 1973 paper Bennett called it a 'logically reversible Turing machine') on which you could write, with a special pen, all the working for a problem, including the final answer. Then, you copy the answer onto a sheet of paper, and retrace all the writing on the tablet, using the pen as an eraser, removing the working as you go. You are left with the answer, and a blank tablet ready to use again.

Incidentally, the same discovery was made at about the same time by Oliver Penrose, a British theorist associated with the Open University; but Penrose's main research interests did not lie in computation and he did not follow up the idea. Bennett, but not Penrose, also made a connection which surely would have delighted Alan Turing, by discussing the biosynthesis of messenger RNA in living cells as an example of reversible computation.

Bennett's discovery was purely theoretical. He had not worked out the practicalities of the kind of gates needed to make such a computer, but he had proved that there was nothing in the laws of physics to forbid it. The next practical step was taken by Ed Fredkin, a scientist with an unusual background, who didn't even know about Bennett's work at the time he made his contribution.

Fredkin was an American college dropout who trained as a fighter pilot in the mid-1950s but had to quit flying because of asthma, and so worked for the air force on a project which introduced him to computer programming. After he returned to civilian life, he became a computer consultant, then founded his own company and became rich, all the while developing the idea, which most people dismissed as crazy at the time, that the Universe might be a gigantic digital computer. One of the reasons why people dismissed the idea as crazy was, of course, that the laws of physics are reversible, and in the 1950s and 1960s it was thought that computers had to be irreversible. But Fredkin was stubborn. He decided that if that was the flaw in the argument, he had to find a way to show that it was possible to make reversible computers. Fredkin had contacts at MIT through his computer company, and managed to wangle an appointment there, in 1966, as a visiting professor. He was so successful that the next year he became a full professor, at the age of thirty-four (still without having finished college!), and then Director of the Laboratory for Computer Science. It was in this capacity that he met Richard Feynman, and arranged to spend 1974 at Caltech so that Feynman could teach him quantum physics and he could teach Feynman about computers.[15] It was while he was at Caltech that Fredkin found the way to make a reversible computer.

Remember the NOT gate? Whichever input it is given, the output is the opposite. Put in 1, you get 0 out. Put in 0, you get 1 out. This is clearly reversible, but doesn't give you much scope to do interesting things. Fredkin's brilliant idea was to invent what is generally referred to as the Fredkin gate, which is complicated enough to do interesting things, but is still reversible.

A Fredkin gate has three channels – three inputs and three outputs. One of the channels, denoted by the letter c, is the control. This always passes the input through the gate unchanged. The other two channels, which we might as well call a and b, also pass the input unchanged if the c input is 0, but swap it over if the c input is 1. So if the input is $c = 1$, $a = 1$ and $b = 0$, the output is $c = 1$, $a = 0$ and $b = 1$. It is pretty obvious that this is reversible; even better, if the output of one Fredkin gate is fed straight into another Fredkin gate (c to c, a to a, b to b), the output is the same as the input to the first gate. Better still, it is possible, though I won't go into details here, to build *any* logic circuit using *only* Fredkin gates, in combinations where, for example, output c from one gate becomes input a for the next gate and output a goes to input c, while output b from the first gate becomes input b for the next gate. The final proof of this was found by one of Fredkin's students, Guy Steel, back at MIT.

All this means that it is in principle possible to build a reversible classical computer, which could simulate precisely the classical laws of physics, and that Fredkin's idea of the Universe being a gigantic computer is not so crazy after all. But at a fundamental level, the Universe operates on the basis of quantum physics, not classical physics. So the next question was, could a computer precisely simulate quantum physics? This is where Feynman comes back into the story.

At the end of the 1970s, Paul Benioff, of the Argonne National Laboratory in Illinois, developed the idea of a Turing machine, operating on a 'tape' in the way Turing described in his classic paper of 1936, which ran on quantum mechanical principles. Although his argument was rather complicated and difficult to follow for non-specialists, he was able to show that it is in principle possible to build a classical computer that operates on quantum principles. Since the real world seems to run on quantum principles, this was an important step forward; but it left open the question of whether a classical (or any) computer could be built that would simulate perfectly the quantum world.

One of the places where Benioff presented his ideas was MIT, where he put them to a meeting in 1981 where Feynman was the keynote speaker. Feynman opened the proceedings with a talk titled 'Simulating Physics with Computers', in which he credited Ed Fredkin with stimulating his interest in the subject, and tackled two questions: Is it possible to simulate physics (meaning quantum physics) with a quantum computer? And, is it possible to simulate (quantum) physics with a classical computer (by simulating 'probability' in some way)?

Curiously, Feynman referred to the discussion of quantum simulators as 'a side-remark', which he dealt with briefly before moving on to the second question, which he regarded as more important. He gave an example of how a 'universal quantum simulator' might work, and said, 'I therefore believe it's true that with a suitable class of quantum machines you could imitate any quantum system, including the physical world,' but was unable to offer a definitive proof that this is the case. He was, though, able to offer a definitive proof that quantum systems can*not* be 'probabilistically simulated by a classical computer'.[16]

The proof depends on the properties of pairs of particles that interact with one another and then fly off in different directions. Like the two-cat version of Schrödinger's puzzle, the behaviour of one particle depends on what happens to the other particle even when they are far apart; this is known as 'entanglement', a term coined by Schrödinger himself. What the argument boils down to is the comparison of a simple pair of numbers, which can be (and have been) measured in experiments. If one number is bigger than the other, there is no way in which quantum mechanics can be simulated perfectly by a classical computer, and therefore no way in which the world can be simulated perfectly by a classical computer. Feynman was delighted by this argument. 'I've entertained myself always,' he said, 'by squeezing the difficulty of quantum mechanics into a smaller and smaller place, so as to get more and more worried about this particular item. It seems to be almost ridiculous that you can squeeze it to a numerical question that one thing is bigger than another. But there you are.'

What Feynman, who was not always scrupulous about giving credit to others, neglected to tell his audience was that the entire argument had been taken, lock stock and barrel, from the work of CERN physicist John Bell, and is usually known as the Bell Inequality. Although Bell himself had not applied his ideas to quantum computers – or quantum simulators, in Feynman's terminology – the idea is so important, both in terms of our understanding of reality and in terms of quantum computation, that it will be the subject of my next chapter. Feynman's conclusion from 1981 is, though, still apposite: 'Nature isn't classical, dammit, and if you want to make a simulation of nature, you'd better make it quantum mechanical.'

Feynman himself did think further about computation in general and about quantum computers. He gave a course on computation at Caltech in the mid-1980s, and a talk in

Anaheim in 1984 in which he described the basis of a quantum mechanical computer,[17] along the lines of Benioff's quantum Turing machine, using reversible gates.[18] In that talk he came up with another of his memorable comments: 'It seems that the laws of physics present no barrier to reducing the size of computers until bits are the size of atoms, and quantum behavior holds dominant sway.' But he never seems to have put two (from his 1981 lecture) and two (from his 1984 lecture) together and realized that such a computer would be fundamentally different from a classical computer not just in terms of its physics but in terms of the kinds of problem it could solve. That leap of inspiration came from Oxford theorist David Deutsch, also in the mid-1980s, and will form the heart of Part Three of this book.

CHAPTER FOUR

Bell and the Tangled Web

The tangled story of entanglement[1] begins, inasmuch as it has a beginning, with the work of a French nobleman, Louis de Broglie, in the 1920s. De Broglie, who held the honorary title of 'Duke', was a latecomer to research in physics. Born in 1892, as the younger son of an aristocratic family he was expected to make a career in the diplomatic service; but under the influence of his elder brother Maurice,[2] who became a physicist in spite of the strenuous objections of their father, he too began to study physics, alongside his 'proper' course in history, at the Sorbonne in 1909. He hoped to move on to research, but his career was interrupted by the First World War, during which he served in the radio communications branch of the army, including a spell based at the Eiffel Tower, which was used as a radio mast. So it was not until 1924, when he was already in his early thirties, that de Broglie was able to submit a thesis for his PhD; but what a thesis!

De Broglie was not a great mathematician but he did have very good physical insight. He was one of the first people to fully accept the idea of light quanta (what we now call photons), and picked up on a curious feature of the equations that Einstein had used to describe these entities. The equations showed a relationship between the wave properties of a photon (its wavelength or frequency) and its particle properties (such as momentum and energy). They showed that light 'waves' could also be treated as particles, and that if you knew the wavelength of a photon you could calculate its momentum. De Broglie pointed out that the same equations worked in reverse – that 'particles' (specifically, electrons) should also behave as waves, if the equations were correct. If you knew the momentum of an electron you could calculate its wavelength. De Broglie's thesis supervisor, Paul Langevin, didn't know what to make of this and showed the work to Einstein, who said, 'I believe that it involves more than a mere analogy'. De Broglie got his PhD, and within three years experiments had been carried out which showed conclusively that electrons do indeed behave as waves exactly in accordance with his description. De Broglie's thesis work, for which he received the Nobel Prize in 1929, was also the inspiration for Erwin Schrödinger's development of the wave version of quantum mechanics.

So by 1927, soon after Schrödinger's wave mechanics and the particle version of quantum mechanics developed by Heisenberg and others had become established, de Broglie should have been a man whose ideas were taken seriously. Instead, his next big idea was first derided and then largely ignored.

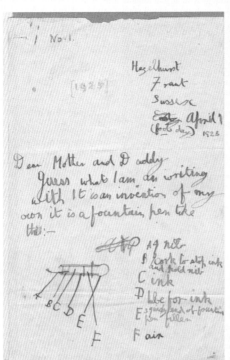

The schoolboy Alan Turing in the mid-1920s (*left*). On the far left is a letter home that he wrote using a fountain pen he had made himself. Twenty years later (*below*) he was a talented long-distance runner.

Bletchley Park. *Below*, Hut 3, where Turing worked on Enigma. *Bottom*, one of the Hut 3 teams at work. *Below right*, two Wrens working with Colossus, the world's first electronic programmable computer.

Left: Two of the American computer pioneers, Presper Eckert (*left foreground*) and John Mauchly (*leaning on pillar*), with the Electronic Numerical Integrator and Computer (ENIAC) at the University of Pennsylvania, around 1946.

Above: Johnny von Neumann (*right*) with Robert Oppenheimer in front of the Institute for Advanced Study computer, 1952.

Right: brochure for the Bendix G-15 computer, 1955. This machine was based on Turing's ACE design.

POWERFUL, LOW COST ... EASY TO USE

Bendix G-15

GENERAL PURPOSE DIGITAL COMPUTER

Above: Astronomer Fred Hoyle with a model 'radio telescope' used in the TV production of his play *A for Andromeda*, which featured Julie Christie (1961).

Below: A version of the 'game of life' in which squares live, die or reproduce according to their relationship with adjacent squares.

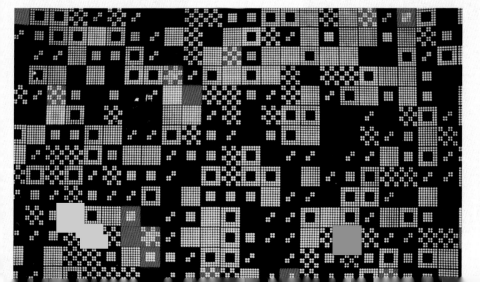

A. PICCARD   E. HENRIOT   P. EHRENFEST   Ed. HERZEN   Th. DE DONDER   E. SCHRÖDINGER   E. VERSCHAFFELT   W. PAULI   W. HEISENBERG   R.H. FOWLER   L. BRILLOUIN

P. DEBYE   M. KNUDSEN   W.L. BRAGG   H.A. KRAMERS   P.A.M. DIRAC   A.H. COMPTON   L. de BROGLIE   M. BORN   N. BOHR

I. LANGMUIR   M. PLANCK   Mme CURIE   H.A. LORENTZ   A. EINSTEIN   P. LANGEVIN   Ch.E. GUYE   C.T.R. WILSON   O.W. RICHARDSON

Absents : Sir W.H. BRAGG, H. DESLANDRES et E. VAN AUBEL

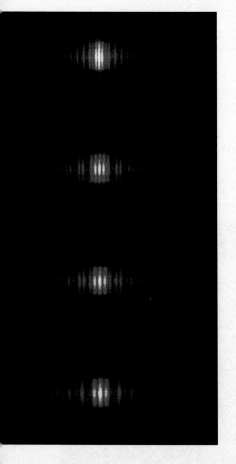

Above: The greatest gathering of physicists ever – Solvay Congress, October 1927. Delegates included quantum pioneers Albert Einstein, Max Planck, Paul Dirac and Erwin Schrödinger.

Left: A set of 'double-slit' diffraction patterns for light.

Below: The full story is in this brief history of quantum physics.

Left: John Bell, who stimulated research that proved the world does not conform to local reality.

Above: David Bohm, who challenged conventional quantum wisdom.

Below: Alain Aspect, who measured the Bell inequalities.

Above left: Hans Dehmelt, who shared the 1989 Nobel Prize in physics for his work with ion traps.

Above right: Brian Josephson celebrates the news of his share of the Nobel Prize in 1973.

Above: David Wineland with an ion trap device.

Left: Wineland and Serge Haroche, the joint winners of the 2012 Nobel Prize in physics.

Left: World chess champion Gary Kasparov ruminates before making his first move in a chess game against the computer Deep Blue, 1997. Deep Blue won the series 3½–2½.

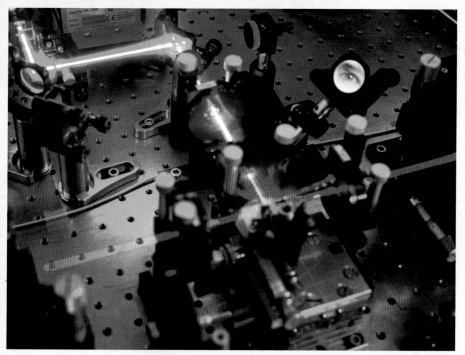

Above: Entanglement apparatus in the lab of Anton Zeilinger, Vienna.

Right: Quantum physics at work: an MRI scanner in operation.

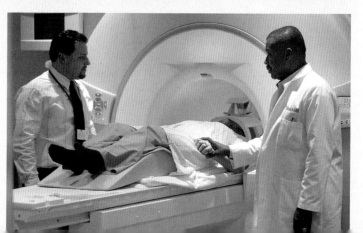

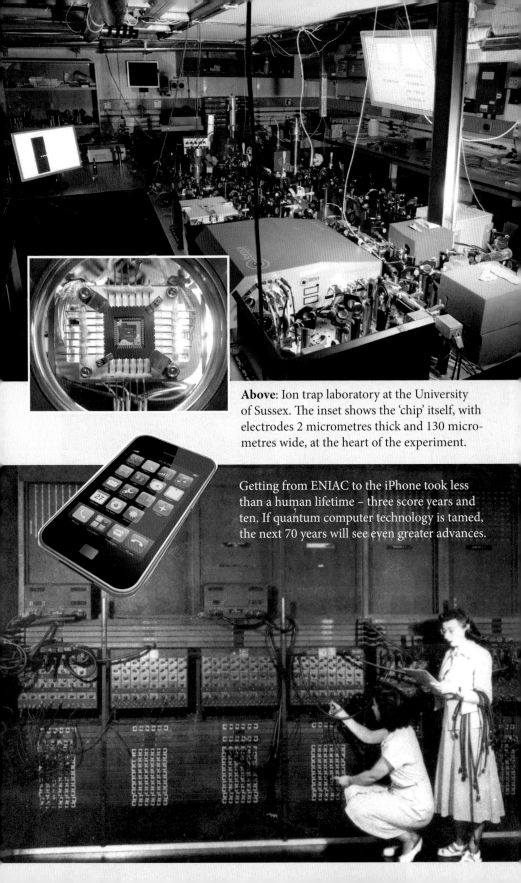

Above: Ion trap laboratory at the University of Sussex. The inset shows the 'chip' itself, with electrodes 2 micrometres thick and 130 micrometres wide, at the heart of the experiment.

Getting from ENIAC to the iPhone took less than a human lifetime – three score years and ten. If quantum computer technology is tamed, the next 70 years will see even greater advances.

Dropping the pilot

The idea was as brilliantly simple as his earlier insight into the nature of wave–particle duality. While other researchers were struggling to choose between the wave version of quantum mechanics and the particle version of quantum mechanics, de Broglie said, in effect, why not have both? He suggested that the wave and the particle were equally real, and that what goes on in experiments such as the experiment with two holes is that a real wave (which became known, for obvious reasons, as a 'pilot' wave) spreads through the apparatus and is responsible both for the interference and for guiding an equally real particle along trajectories determined by the interference, like a surfer riding the waves in the sea. The key point on this picture is that the particle always exists at a definite place even when it is not being measured or observed. Equally, the apparatus (in principle, the whole world) is occupied by a physically real field, like the electromagnetic field, which we are only aware of because of its influence on the statistical behaviour of particles. Electrons (or photons) fired one after another through the experiment with two holes do not all follow exactly the same trajectory because of tiny differences in the speed or direction with which they start out. We can never measure the field, but we can measure the particles. This is known as a 'hidden variables' theory, although as a result of what John Bell called 'historical silliness'[3] it is the things we *can* measure, such as the positions of particles, that are regarded as the 'hidden' variables, while the thing we *can't* measure, the field, is not. As Bell also said, 'this idea seems so natural and simple, to resolve the wave–particle dilemma in such a clear and ordinary way, that it is a great mystery to me that it was so generally ignored'.[4]

But ignored it was. De Broglie presented his ideas (in a much more fully worked out form than the sketch I have provided here) to the same Solvay Congress in 1927 that heard Schrödinger point out that there is nothing in the equations to suggest that the wave function 'collapses', as proposed by Niels Bohr and followers of his 'Copenhagen Interpretation'. De Broglie's pilot wave idea was savagely attacked on mathematical grounds by Wolfgang Pauli, a mathematical physicist with a (largely justified) high opinion of his own ability and a (not always justified) tendency to dismiss the efforts of those he regarded as lesser minds. Although de Broglie did respond to these criticisms, and a modern reading of the proceedings of the meeting suggests that he successfully answered them, his diffident style left the impression that he had lost the argument. In addition, Pauli pointed out that the hidden variables idea led to some peculiar implications when applied to more than one particle. Crushed, de Broglie gave up promoting his idea. Those peculiar implications concern entanglement and what Einstein would later call 'spooky action at a distance', and are at the heart of the modern understanding of quantum physics and quantum computation. But what seemed at the time the final nail in the coffin of de Broglie's idea came in 1932, when Johnny von Neumann published his great book on quantum mechanics, in which, among other things, he 'proved' that no hidden variables theory could be an accurate description of reality. He was wrong.

Von Neumann gets it wrong

In order to appreciate just how wrong von Neumann was, we need to be clear about what hidden variables theories are –

and a good way to see that is by comparison with the all-conquering (for half a century) Copenhagen Interpretation. There are three significant points of conflict between these two views of the world:

1   The Copenhagen Interpretation says that the wave function is a complete description of a system and contains all the information about it. Hidden variables theory says that the wave function is only part of the story, and that there are also particles with real positions and momenta.

2   The Copenhagen Interpretation says that the wave mostly spreads out in line with Schrödinger's equation, but sometimes 'collapses' to (more or less) a point. The mechanism for this collapse has never been satisfactorily explained. Hidden variables theory says that the wave always develops in line with Schrödinger's equation; there is no collapse.

3   The Copenhagen Interpretation says that even though the evolution of the wave is deterministic, the process of collapse introduces an element of probability into the outcomes of experiments – that is, that quantum physics is stochastic. Hidden variables theory says that everything is deterministic, and that the only reason we cannot predict everything perfectly is that we can never know perfectly all of the starting conditions.

Stated like that, it is hard to see how any sensible person could choose the Copenhagen Interpretation over hidden variables theory unless there were very powerful evidence against the latter. But the power of von Neumann's 'proof' lay as much in his name as in the mathematics; not only that, it was refuted almost as soon as it was published, but the refutation was ignored. It is probably significant that the flaw in von Neumann's argument was found by a young researcher who came from outside the quantum mechanical fold, and was unlikely to be overawed by von Neumann's (or Bohr's) reputation; but the fact that she was a young outsider is also one factor in explaining why her refutation was ignored.

Grete Hermann was born in Bremen in 1901. She qualified as a secondary school teacher in 1921, then studied mathematics and philosophy, working under Emmy Noether in Göttingen and receiving a PhD in 1926 for research into what would later become known as computer algebra. For some years she stayed on in Göttingen, developing her interest in philosophy and the foundations of science, but she was also a committed socialist, editing a newspaper called *The Spark* and actively opposing the increasingly powerful Nazis. In 1936 she had to flee Germany to escape the regime, travelling via Denmark and France to England. In 1946 she returned to Germany and worked in education, helping to rebuild post-war society; she died in 1984, having lived long enough to see her early insight come in from the cold.

In 1934, Hermann had visited Werner Heisenberg's group in Leipzig to discuss the foundations of quantum physics in relation to her favoured Kantian philosophy. It was there that she confronted von Neumann's 'proof' of the impossibility of hidden variables theories, and pointed the flaws out to Heisenberg and his colleagues. The problem was not with the mathematics – von Neumann was not prone to making mathematical mistakes – but with the assumptions that went into the mathematics. Fortunately, this means that we do not need to go through the maths to understand what went wrong; but, surprisingly, it means that the mistake ought to have been obvious to anyone who looked carefully, as Hermann did, at what von Neumann was saying. His mistake involved an assumption about the way averages are taken in quantum mechanics. There is no perfect way of explaining the error in everyday language, but one that may help is to look at the way we take averages in everyday life. Imagine that we have two groups of people. In one group there are ten people, and their average height is 1.8 metres. In the other group there are twenty people, and their average height is 1.5 metres. Von Neumann's mistake was equivalent to saying that therefore the average height of the thirty people is 1.65 metres, the average of 1.8 and 1.5. But because there are twice as many people in the second group, the correct average is, of course, 1.6 metres.

Hermann's discovery of the failure of von Neumann's 'proof' was published in 1935, but only in German and in a journal devoted more to philosophy than to physics, which physicists did not read. That also partly explains why it faded into obscurity.[5] But Heisenberg knew of it, directly from Hermann, and it is a mystery that will now never be resolved why he did not at least draw it to the attention of a wider circle of quantum physicists. Hermann herself was never in a position to promote her finding, even if she had wanted to, because of the dramatic changes in her life and career brought about by the political developments in Germany; in any case, though, she does not seem to have been particularly enamoured of hidden variables theories, and had simply been concerned as a mathematician with setting the record straight about von Neumann's claim. To all intents and purposes, in the quantum physics community it was as if Hermann had not existed, and from 1932 onwards 'everybody knew' that hidden variables theories would not work, because Johnny von Neumann said so. Even people who never read his book believed this, because they had been told it was so. But while hidden variables theories languished, Einstein, along with a couple of colleagues, pointed out another problem with the Copenhagen Interpretation – in fact, an aspect of the 'problem' with de Broglie's pilot wave pointed out by Pauli.

Spooky action at a distance

Einstein first expressed the idea behind what became known as the 'EPR paradox', although I prefer the term 'puzzle', in 1933, at another Solvay Congress. Leon Rosenfeld, one of Niels Bohr's collaborators, recalled that after a lecture by Bohr Einstein pointed out to Rosenfeld that if two particles interacted with one another and then flew apart,

> an observer who gets hold of one of the particles, far away from the region of interaction, and measures its momentum ... [will] from the conditions of the experiment, ... obviously be able to deduce the momentum of the other particle. If, however, he chooses to measure the position of the first particle, he will be able to tell where the other particle is ... is it not paradoxical? How can the final state of the second particle be influenced by a measurement performed on the first, after all physical connection has ceased between them?[6]

His point is that, according to the Copenhagen Interpretation, neither particle has a definite position or momentum until it is measured. So if measuring one particle immediately determines the state of the other particle, the particles seem to be connected by what he later called 'spooky action at a distance',[7] as in my 'thought experiment' involving the two cats. Rosenfeld said later that at the time he didn't realize that Einstein saw this as anything more than an illustration of the unfamiliar features of quantum phenomena, and that two years later, when Einstein presented a more complete version of the puzzle in a paper co-authored with Boris Podolsky and Nathan Rosen (hence EPR), it came to Bohr's group 'as a bolt from the blue'.[8]

Shortly after the 1933 Solvay Congress, Einstein, like Hermann and so many other scientists a refugee from Nazi persecution, settled in what became his final academic home, the Institute for Advanced Study in Princeton. He was assigned Nathan Rosen, then aged twenty-four, as his assistant, and he already knew Boris Podolsky, a Russian-born American physicist who had moved to the United States with his family in 1913, when he was seventeen. The basic idea of the EPR paper was, of course, Einstein's; Rosen helped as a good assistant should, and Podolsky was recruited not only as a sounding board but to write the paper, since at that time Einstein's English was rather limited. Indeed, Podolsky's English wasn't perfect, as the missing 'the' in the title of the paper, 'Can Quantum-Mechanical Description of Physical Reality be Considered Complete?', suggests. More seriously, though, Einstein was never happy with the published version of the paper, feeling that Podolsky had obscured his fundamental point by elaborating the argument along philosophical lines. That need not bother us, since the essence is what Einstein said to Rosenfeld in 1933. But a couple of quotes from the EPR paper are apposite. First, a definition of an 'element of physical reality', with their emphasis: '*If, without in any way disturbing a system, we can predict with certainty (i.e., with probability equal to unity) the value of a physical quantity, then there exists an element of physical reality corresponding to this physical quantity.*' In other words, particles *do* have real positions and momenta, even when we are not observing them.[9] Secondly, the paper points out that the Copenhagen Interpretation 'makes the reality of P and Q [momentum and position for the second particle] depend on the process of measurement carried out on the first system, which does not disturb the second system in any way. No reasonable definition of reality could be expected to permit this.' So: 'We are thus forced to the conclusion that the quantum-mechanical description of physical reality given by the wave function is not complete.' As we shall see, Einstein was right on this point; but, ironically, he was (probably) wrong about spooky action at a distance, and the 'reasonableness' of reality.

Schrödinger responded enthusiastically to the paper, writing to Einstein that 'my interpretation is that we do not have a q.m. that is consistent with relativity theory, i.e., with a finite transmission speed of all influences'. But he, too, was right for the wrong reason.

The most important point is made right at the end of the paper: 'While we have thus shown that the wave function does not provide a complete description of the physical reality, we left open the question of whether or not such a description exists. We believe, however, that such a theory is possible.' Not only was it possible – de Broglie had already found it! But de Broglie's description of reality also includes the spooky action at a distance that Einstein abhorred: for the model that de Broglie had found, when applied to two or more particles, also requires that they influence one another in this way. This property is known more formally, as I have already noted in passing, as non-locality.[10] In 1935, though, nobody worried about that. After all, von Neumann had 'proved' that hidden variables theories were impossible. Instead, the EPR paper provoked a frantic burst of work by Bohr and his colleagues, culminating in a paper claiming to refute the argument. Bohr's counter-argument is itself rather woolly and 'philosophical' in the sense Einstein disliked, and didn't really resolve the debate, so I shall not go into it here. A few people puzzled about the meaning of the EPR paper and the nature of reality, but most physicists were happy to accept the Copenhagen Interpretation FAPP, and very soon the outbreak of the Second World War gave them much more immediate problems to tackle. It was not until things settled down after the war that somebody did what was thought by all those who trusted von Neumann to be impossible and, without knowing about de Broglie's work, came up with a working hidden variables theory.

Bohm does the impossible

David Bohm didn't start out trying to upset the quantum mechanical applecart. He had been born in 1917 in the relative backwater of Wilkes-Barre, Pennsylvania, where his early interest in science was stirred by reading science fiction. He graduated from Pennsylvania State University in 1939, and moved on to Caltech and then to the University of California, Berkeley, to work for a PhD under Robert Oppenheimer. As a student, he was active in left-wing politics and a member of the Young Communist League. When Oppenheimer was given responsibility for the scientific side of the Manhattan Project to develop a nuclear bomb, he wanted Bohm to come with him to the new secret research laboratory at Los Alamos, but the military authorities vetoed this because of Bohm's political affiliations. So Bohm stayed in Berkeley, but worked on a problem involving interactions between protons and neutrons that was relevant to the Manhattan Project, reporting his results to Oppenheimer. Because of the secrecy surrounding this project, when he completed this work in 1943 he was not even allowed to write up a formal PhD thesis, let alone explain his work to the examiners, who awarded the degree solely on Oppenheimer's word that the work merited it. He continued to do theoretical work relevant to the development of the nuclear bomb until the end of the war, then moved to Princeton University.

At Princeton, after teaching a course on quantum mechanics Bohm wrote a textbook on the subject (entitled simply *Quantum Theory*) that was published in 1951 and became an instant classic. He wrote it 'primarily in order to obtain a better understanding of the subject'. It set out clearly the standard Copenhagen Interpretation, and led to discussions with Einstein, who had seen an early version, and who said that while Bohm had 'explained Bohr's point of view as well as could probably be done', that didn't mean Bohr was right. Bohm himself reflected: 'After the work was finished, I looked back over what I had done and still felt somewhat dissatisfied.'[11] But in the book, as well as explaining Bohr's version of quantum mechanics better than Bohr did himself, Bohm also explains the EPR puzzle better than Podolsky had done in the EPR paper itself, using the example of particle spin rather than position and momentum. This is the way the puzzle has usually been described since that time, and the way the predictions were tested in experiments, so it is worth elaborating a little.

In this version of the experiment (still purely hypothetical in 1951), two particles interact and fly off in different directions as before. But the important characteristic of the particles now is that they must have opposite spins to one another. 'Roughly speaking,' said Bohm, 'this means that the spin of each particle points in a direction exactly opposite to that of the other, insofar as the spin may be said to have any definite direction at all.' But, according to the Copenhagen Interpretation, the direction of spin is not definite – it is not an 'element of physical reality' – until it is measured. It could be up, down, sideways, or at any angle in between. And a spin at any angle can be specified in terms of a combination of a certain amount of 'up' spin and a certain amount of 'sideways' spin (these are known as the spin components). In the experiment imagined by Bohm, an experimenter can choose which direction of spin (which component) to measure *after* the two particles have interacted. Whichever component is measured, the equivalent component of the other particle must have the opposite spin. How is this possible unless the spins were indeed 'elements of physical reality' (hidden variables) all along?

Having presented this clear explanation of the EPR puzzle, Bohm proceeded to knock it down! He claimed that Einstein and his colleagues had made an invalid assumption (since he was wrong, I won't go into details) and concluded that 'no theory of mechanically determined hidden variables can lead to *all* of the results of the quantum theory [by which he means the Copenhagen Interpretation].' He acknowledged that it might in principle be possible to test this assertion by experiment, and that although 'unfortunately, such an experiment is still far beyond present techniques . . . it is quite possible that it could some day be carried out. Until and unless some such disagreement between quantum theory and experiment is found however, it seems wisest to assume that quantum theory is substantially correct.'

But even while the book was going through the publishing process, Bohm changed his mind, partly as a result of discussions with Einstein. He found an 'impossible' hidden variables theory that worked (essentially, a rediscovery in more fully worked out form of de Broglie's pilot wave), and published a pair of papers on the subject in the prestigious *Physical Review* in 1952. But by then Bohm's life was in turmoil, and he was in no position to promote the idea effectively.

At the end of the 1940s there was a growing paranoia in America about the 'communist threat', and as the Cold War began Congress set up the now notorious House Un-American Affairs Committee, chaired by Senator Joseph McCarthy, to seek out communist sympathizers in positions of influence or responsibility. The nature of their investigations is neatly summed up by the present-day definition of 'McCarthyism' in Wikipedia: the practice of making accusations of disloyalty, subversion or treason without proper regard for evidence. As somebody who had worked, if only peripherally, on the Manhattan Project, and had once been a member of a communist organization, Bohm was an early target of the committee; he was called before it in the spring of 1949, but refused to testify. As a result, he was charged with contempt of Congress, and suspended by Princeton University. Even though he was acquitted in 1951, and according to legend Einstein asked for Bohm to return as his assistant, Princeton refused to renew his contract. With no prospect of continuing his career in the United States, Bohm moved to Brazil, and was based at the University of São Paulo when his two-part paper on 'A Suggested Interpretation of the Quantum Theory in terms of "Hidden" Variables' appeared in the *Physical Review* in 1952. Never happy in Brazil, Bohm moved to Israel in 1957, then to England, eventually (in 1961) settling at Birkbeck College in London. He died in 1992.

Bohm's big idea met with an unenthusiastic response. The old guard of quantum physicists – people like Heisenberg and Pauli – were too set in their ways and dismissed it out of hand. The younger generation, raised on the Copenhagen Interpretation and happy that it worked FAPP, could not be bothered with a new idea that 'only', as far as they could see, reproduced all the results of the Copenhagen Interpretation. To such people, Bohm's papers did not seem to contain any new physics. Just two people were impressed: Louis de Broglie, who happily took up this revival of the pilot wave idea; and a young physicist working at the UK Atomic Energy Research Establishment, who knew about de Broglie's work and von Neumann's 'proof', and was astonished 'to see the impossible done'. His name was John Bell, and he would-be responsible for the most profound re-interpretation of quantum mechanics since its inception.

From Belfast to Bohm, and beyond

John Stewart Bell was born in Belfast on 28 July 1928 – the year after de Broglie presented his pilot wave idea to an un-impressed Solvay Congress. Bell's father was also named

John, so 'our' John Bell was known as Stewart to his family. In an interview with Jeremy Bernstein, Bell described his family as 'poor but honest', traditional working-class people typical of Northern Ireland at the time.[12] His father had worked as a horse dealer, village blacksmith, and self-taught electrician; with an older daughter, Ruby, and two younger sons, David and Robert, to support as well as John, the family finances were stretched to the limit, and the usual expectation in families such as these was the children would leave school at about the age of fourteen and begin working. But, like many other working-class mothers, John senior's wife, Annie, saw education as the only way out of poverty. John Stewart was a bright child with an early interest in books and science, and did very well at his first schools, Ulsterville Avenue and Fane Street. When he was eleven he easily passed the examination allowing him to move on to secondary school; but secondary education in the UK was not free at the time, so the only hope was for the boy to pass more examinations which would be rewarded with scholarships. Annie encouraged his ambition to move on to secondary school, and although 'I sat many examinations for the more prestigious secondary schools, hoping for scholarships, but I didn't win any,' she found funding from somewhere – John never knew where – for him to attend Belfast Technical High School, the least expensive option, where alongside the usual academic courses he was taught bricklaying, carpentry and book-keeping.

Bell did not, though, confine himself to the curriculum. Although he had been confirmed in the Protestant Church of Ireland, like many adolescents he began to question the teaching of the Church, and to doubt the existence of God. Unlike many adolescents, he sought answers to these questions by reading 'thick books on Greek philosophy', but soon became disillusioned with philosophy as well, and turned to physics. 'Although physics does not address itself to the "biggest" questions, still it does try to find out what the world is like. And it progresses. One generation builds on the work of another.' Bell's own progress then took a providential detour. At the age of sixteen, he graduated from the High School and qualified academically to move on to university. The only financially viable option was for him to carry on living with his parents and attend Queen's University in Belfast. This was (and is) an excellent university, but would not admit anyone younger than seventeen. In order to fill in the year he would have to wait, Bell applied unsuccessfully for many jobs before ending up as a laboratory assistant at the university itself. It was the best thing that could have happened. He met his future professors, who took a keen interest in this highly motivated young man, lent him books and gave him advice. In effect, 'I did the first year of my college physics when I was cleaning out the lab and setting out wires for the students.' With this flying start, when Bell finally began his undergraduate studies in 1945 he sailed through the course, graduating with first class honours in experimental physics in 1948. By then, he had become intrigued by quantum physics (indeed, he had read popular books on quantum mechanics when he was still at the High School), and was able to stay on for a year, earning a first class degree in mathematical physics in 1949. He had also read Max Born's book *Natural Philosophy of Cause and Chance*, and accepted at face value Born's assertion that von Neumann had proved hidden variables theories to be impossible. Referring to von Neumann's 'brilliant book', Born said:

No concealed parameters can be introduced with the help of which the indeterministic description [of quantum physics] could be transformed into a deterministic one. Hence if a future theory should be deterministic, it cannot be a modification of the present one but must be essentially different. How this could be possible without sacrificing a whole treasury of well-established results I leave to the determinists to worry about.

Bell would have liked to make a career in research, and his professor at Queen's, Peter Paul Ewald (one of the many scientists to have fled Nazi persecution in the 1930s), suggested that he should join Rudolf Peierls' group in Birmingham as a PhD student. This was Bell's dream; but by 1949, at the age of twenty-one, he was feeling guilty about having lived off his parents for so long, and knew that he would need help to follow up this opportunity. He felt it his duty to get a job, and found one with the UK Atomic Energy Research Establishment at Harwell, in Oxfordshire, soon moving to join the AERE's particle accelerator design team based at Malvern, Worcestershire, where his practical background and deep understanding of physics were put to good use.

At Malvern, Bell met a young Scottish physicist, Mary Ross, from Glasgow. She came from a slightly more affluent family than the Bells, and had had the good fortune of attending a co-educational school which covered both the 'primary' and 'secondary' years. Because there were boys as well as girls at the school, there was a physics course; at that time it would have been most unlikely for a girls-only school to teach physics. Mary did well at school, especially in physics, and in 1941 went on to the University of Glasgow. Her education was interrupted by compulsory war work in the radar lab at Malvern, which she hated. But after the war she completed her studies and returned to Malvern in a much more congenial role as a member of the accelerator design group. John and Mary married in 1954, and stayed happily together until his death.

In 1952, Bell became a consultant with the British team contributing to the design of what would become the first particle accelerator at the new European research centre in Switzerland, CERN. The machine, which started operating in 1959, was a proton synchrotron which, at the time, was the world's highest-energy particle accelerator. But something else, far more significant, also happened in 1952. Bell was chosen by AERE to become (in 1953) one of the beneficiaries of a scheme whereby some of their young scientists were given leave to spend a year at university carrying out research. It wasn't something you applied for, and Bell always expressed astonishment that he had been selected, as he saw it, out of the blue. It must rank as one of the most important decisions in the history of science. He went to Birmingham, and 'there I became a quantum field theorist'.

Rudolf Peierls suggested an area of research for Bell to work in, and this led to his developing an important piece of quantum field theory known as the CPT theorem. Unfortunately for Bell, while he was writing up this work for publication he found that Gerhard Lüders, based at the University of Göttingen, had just published a paper announcing the same discovery. His own work was, nevertheless, a significant piece of research, and combined with another completed after he returned to Harwell it formed the basis of a thesis for which Bell was awarded a PhD in 1956.

It was also in 1952 that Bell 'saw the impossible done', as he put it in his address to the de Broglie 90th birthday symposium thirty years later. David Bohm showed how 'the indeterministic description [of quantum mechanics] could be transformed into a deterministic one'. This set questions buzzing in Bell's brain: 'But why then had Born not told me of this "pilot wave"? If only to point out what was wrong with it? Why did von Neumann not consider it? . . . Why is the pilot wave picture ignored in text books?' Bell saw that von Neumann must have been wrong, and discussed the puzzle with a German colleague, Franz Mandl, who was able to give him the gist of von Neumann's argument. As Bell told Bernstein, 'I already felt that I saw what von Neumann's unreasonable axiom was.' He also commented: 'I hesitated to think [quantum mechanics] might be wrong, but I *knew* that it was rotten.' Even so, a full-scale assault on the puzzle would have to wait, since Bell had returned to Harwell to join a new group, the Theoretical Physics Division, carrying out fundamental research in areas related to particle physics.

Bell had a tenured post at Harwell – a job until he retired and a pension afterwards – but both he and Mary became increasingly unhappy there in the late 1950s. Harwell had been set up to develop peaceful uses of nuclear energy, and by then the first nuclear power stations had been built and future developments were more in the realm of industrial applications than fundamental science. So in 1960 they gave up the security of Harwell and moved to CERN – the ideal place to combine Mary's interest in accelerator design with John's interest in fundamental physics, but with no promise of long-term security. The practice at CERN was to offer a three-year contract, on the understanding that in normal circumstances a second three-year contract would follow; and during those six years, a lucky few would be offered tenured posts. The point was to give as many people as possible a chance to spend some time at the multinational research centre. The Bells would be among the lucky (or rather, talented) few who got the offer of tenure, and stayed there for the rest of their careers.

Although Bell carried out a great deal of research during his time at CERN, and published many important papers, he will be remembered above all for two pieces of work completed while he was in the United States in the mid-1960s. But although this burst of creative activity came to fruition in America, it was undoubtedly stimulated in CERN, in 1963, when a visitor from the nearby University of Geneva, Josef-Maria Jauch, gave a seminar in which he claimed to have 'strengthened' von Neumann's impossibility theorem. Since Bell had already seen the impossible done, this was like the proverbial red rag to a bull, and he determined to resolve the issue once and for all. He was helped by two factors: von Neumann's book was now available in English, so he could examine the 'proof' at first hand; and he and Mary were planning a sabbatical in the United States, where they would visit the Stanford Linear Accelerator Center in California (arriving in November 1963, just after the assassination of John F. Kennedy), Brandeis University in Massachusetts and the University of Wisconsin-Madison. Freed from his regular routine at CERN, Bell had time both to think and to put his thoughts down on paper.

Von Neumann's silly mistake and Bell's inequality

Among the great virtues of the two scientific papers that emerged from Bell's trip to the United States were their clarity and simplicity, related to the fact that with his practical experience in particle physics Bell was able to spell out the kinds of experiments that could, in principle, be carried out to test the ideas – not that he expected, at the time, to see such experiments carried out. The first of the two papers that he wrote (but not the first to be published, as we shall see) was 'On the Problem of Hidden Variables in Quantum Mechanics', in which he analysed the flaws in von Neumann's argument, in the 'refinement' proposed by Jauch, and in a third variation on the theme. At the time, he was quite unaware of Grete Hermann's earlier work. Bell went further than Hermann, though, in not just finding the flaw in von Neumann's argument, but also (actually, at the beginning of his paper) presenting his own version of a hidden variables theory, much simpler than Bohm's model but demonstrating with equal force that the 'impossible' could be done. He also made clear that non-locality ('spooky action at a distance') was an integral part both of Bohm's model and of his own. As Bell commented in the paper, this means that these hidden variables theories (and, he suspected, *all* such theories) resolve the EPR puzzle in exactly the way Einstein would have liked least!

Bell completed 'On the Problem of Hidden Variables in Quantum Mechanics' while at Stanford (although he mentions in the acknowledgements that 'the first ideas of this paper were conceived in 1952'), and sent it off to the journal *Reviews of Modern Physics* in 1964. As usual, the editor of the journal sent the paper to an expert referee to assess its suitability for publication. The referee was sympathetic (it is widely thought that it was Bohm), but suggested some improvements be made before it was accepted for publication. Bell, as authors often do, made the bare minimum of changes to meet the referee's demands, and sent the paper back to the journal. Unfortunately, the revised paper was misfiled, and some time later, thinking it had not come back, the editor wrote to Bell asking where it was. The letter went to Stanford, but by then Bell was back in England. By the time the confusion was sorted out, more than a year had passed and the paper was eventually published in 1966, after the second of Bell's great papers, the one which *proved* that all hidden variables theories must be non-local.

Although there is no need here to go into the details of Bell's refutation of von Neumann's argument, which is essentially the same as Hermann's, it does seem worth re-iterating in Bell's own words (from an interview published in the science and science fiction magazine *Omni* in May 1988) how bizarre it is that people ever took it seriously: 'The von Neumann proof, if you actually come to grips with it, falls apart in your hands! There is *nothing* to it. It's not just flawed, it's *silly*! . . . You may quote me on that: The proof of von Neumann is not merely false but *foolish*!' Indeed, David Mermin, of Cornell University, commented in *Reviews of Modern Physics* itself, in 1993, that the argument is so silly that 'one is led to wonder whether the proof was ever studied by either the students or those who appealed to it to rescue them from speculative adventures [in the realms of quantum inter-pretation]'. Bell's second great paper was scarcely speculative, but it was certainly adventurous.

That second paper was entitled 'On the Einstein–Podolsky–Rosen Paradox', and begins by noting that the EPR argument was advanced in support of the idea that 'quantum mechanics could not be a complete theory but should be supplemented by additional variables. These additional variables were to restore to the theory causality and locality.' He goes on to say: 'in this note that idea will be formulated mathematically and shown to be incompatible with the statistical predictions of quantum mechanics. *It is the requirement of locality, or more precisely that the result of a measurement on one system be unaffected by operations on a distant system with which it has interacted in the past, that creates the essential difficulty.*'[13] In other words, if there is a real world out there independent of our observations (if the Moon is there when nobody is looking at it), then the world is non-local. Equally, though, if you insist on locality, then you have to give up the idea of reality and accept the literal truth of the 'collapse of the wave function' as envisaged by the Copenhagen Interpretation. But you cannot have both – you cannot have local reality.

But the most dramatic feature of Bell's discovery is often overlooked, even today. This is not a result that applies only in the context of quantum mechanics, or a particular version of quantum mechanics, such as the Copenhagen Interpretation or the Many Worlds Interpretation. It applies to the Universe independently of the theory being used to describe the Universe. It is a fundamental property of the Universe not to exhibit local reality.

I do not intend to go into the details of Bell's calculation, which can be found in a thorough but accessible presentation by David Mermin in his book *Boojums All the Way Through*.[14]

It happens that Mermin presents these ideas within the framework of the Copenhagen Interpretation, accepting locality but denying a reality independent of the observer; my own preference is to accept reality and live with non-locality, but this just emphasizes the point that whichever interpretation you use, Bell's result still stands. The crucial point is this: Bell found that if a series of measurements of the spins of particles in a Bohm-type version of the EPR experiment is carried out, with various orientations of the detectors used to measure the spin, then if the world is both real and local the results of one set of measurements will be larger than the results of another set of measurements. This is Bell's inequality. If Bell's inequality is violated, if the results of the second set of measurements are larger than those of the first, it proves that the world does not obey local reality. He then showed that the equations of quantum mechanics tell us that the inequality must indeed be violated. Since then, other similar inequalities have been discovered; all are known as Bell inequalities, even though he did not discover them all himself. The whole package of ideas is known as Bell's theorem.

Bell's conclusion is worth quoting:

> In a theory in which parameters are added to quantum mechanics to determine the results of individual measurements without changing the statistical predictions, there must be a mechanism whereby the setting of one measuring device can influence the reading of another instrument, however remote. Moreover, the signal involved must propagate instantaneously.

It's noteworthy that Bell did not expect to reach such a conclusion when he started out down this path. His instinct was to side with Einstein and assert that local reality must be the basis on which the world works. As he later wrote to the American physicist Nick Herbert:

> I was deeply impressed by Einstein's reservations about quantum mechanics and his views of it as an incomplete theory. For several reasons the time was ripe for me to tackle the problem head on. The result was the reverse of what I had hoped. But I was delighted – in a region of wooliness and obscurity to have come upon something hard and clear.

In the words that Arthur Conan Doyle put into the mouth of Sherlock Holmes in *The Sign of Four*, Bell had eliminated the impossible – local reality – and so what was left, however improbable, had to be the truth.

But it is one thing to prove mathematically that the world is either unreal or non-local, quite another to prove it by experiment. Bell realized this, commenting at the end of his paper: 'The example considered above has the advantage that it requires little imagination to envisage the measurements involved actually being made.' Little imagination, but a great deal of experimental skill. Astonishingly, it was less than ten years before the first such experiments were carried out – and it might have been sooner had Bell's paper not disappeared into a kind of publishing black hole.

Unlike the first of his two great papers, it was printed fairly quickly, in 1964. But, also unlike the first paper, it did not appear in a widely read or prestigious journal, largely because of Bell's reluctance, as a guest at various American research centres, to impose on his hosts by incurring the 'page charges' applied by the more prestigious journals – a fee for publication based on the number of printed pages occupied by the paper.

Bell did most of the work on the paper while at Brandeis, and completed it in Madison. As he explained to Paul Davies: 'Probably I got that equation into my head and out on to paper within about one weekend. But in the previous weeks I had been thinking intensely all around these questions. And in the previous years it had been at the back of my head continually.'[15] Bell chose to send the fruits of all that thinking to a completely new journal, *Physics*, which had no page charges – in fact, it actually gave contributors a small payment for their papers; but this was no real benefit, since in those pre-internet days the contributors in turn had to pay the journal for copies of the paper (reprints) to send to friends and colleagues. The two payments more or less cancelled each other out. It seems that Bell's paper was accepted for publication (in the very first volume of *Physics*) because the editors mistakenly thought that it refuted Bohm's hidden variables interpretation of quantum mechanics.[16] Bell's paper did not make a big splash in 1964. *Physics* was not a widely read journal, and was closed down after only four years. Some of the people who did read the paper probably misunderstood it in the same way that the editors had. But the message got through to a tiny number of researchers, who ended up collaborating and competing in the first experiments to test Bell's theorem.

First fruits

While Bell was at Brandeis, he gave a talk about his work and distributed a few copies (preprints) of his second paper, which had not yet been published. These had a rather unprepossessing appearance, produced by a pre-photocopier duplicating process in smudgy purple ink. At first sight, they looked more like the work of a crank than that of a respectable physicist; but one of these smudgy preprints would have a big impact.

Somebody – just who is lost in the mists of time – sent one of the preprints to Abner Shimony, a physicist working at Boston University. But Shimony was not your average physicist. Born in 1928, the same year as Bell, he had graduated from Yale in 1948 with a combined major in philosophy and mathematics, and received his PhD in philosophy from Yale in 1953. Like Bell, he had turned to philosophy to seek the answers to the big questions about life, the Universe and everything; rather later than Bell, but influenced by Born's classic book, he decided that physics was more likely to provide those answers, and after two years of compulsory military service, in 1955 he embarked on a PhD in physics at Princeton. His time in the army, based in a mathematics section where one of the things he did was teach a course on information theory, was very valuable, Shimony recalls, because it gave him time to read up on under-graduate physics.

At Princeton, one of the first things his supervisor did was to tell him to '"read the paper by Einstein, Podolsky and Rosen on an argument for hidden variables, and find out what's wrong with the argument." So that was my first reading of the EPR paper, and I didn't think anything was wrong with the argument. It seemed to be a very good argument. I never saw anything wrong with it.'[17]

Alongside his physics research, in 1959 Shimony joined the philosophy faculty of MIT. He lectured there on, among other things, the foundations of quantum physics. As a part-time physicist he did not complete his second PhD until 1962, after which he took up a joint appointment in physics and philosophy at Boston University. A couple of years later, he received Bell's preprint out of the blue and, resisting the temptation to throw the scruffy document straight into the waste paper bin, read enough of it to realize its importance, then settled down to take a more detailed look. 'The more I read it, the more brilliant it seemed.' Already familiar with the EPR argument, he was most impressed by the suggestion that these ideas could be tested by experiment.[18] What is more, he already knew of an experiment that had been carried out along these lines, and might be adapted to test Bell's theorem.

In a paper published in 1957, David Bohm and his student Yakir Aharonov had discussed entanglement and drawn attention to an experiment carried out back in 1950, for a completely different reason, which seemed to show entanglement at work. That experiment had been carried out by Chien-Shiung Wu and her student Irving Shaknov, and involved monitoring gamma rays (very high-energy photons, even more energetic than X-rays) produced when an electron meets a positron and annihilates. The relevant property of photons that is measured in such experiments is their polarization, which is analogous to the spin of an electron. The relevant point is that a photon can be polarized in any direction across its line of flight, like the baton twirled by a majorette. The Wu–Shaknov data suggested a correlation between the polarizations of separated photons (implying entanglement), but were not conclusive – which was hardly surprising, since the experiment had not been set up to measure such things.

Shimony did not follow up the implications immediately, but a little later he was visited by a graduate student, Michael Horne, looking for a problem to work on for his PhD. Shimony showed Horne Bell's paper, along with those of Bohm and Aharanov, and Wu and Shaknov, and suggested that he might try to design an experiment to test Bell's theorem. As a preliminary, Horne quickly found that he could construct a simple hidden variables model that would account for the Wu–Shaknov results, but that some much more sophisticated experiment would be needed to provide a proper measurement of Bell's inequality, using polarization measurements at a variety of different angles on pairs of photons produced by a single source. The good news, though, was that you didn't need gamma rays to do the job – ordinary photons of visible light would suffice. Indeed, it is easier to measure the polarization of such 'ordinary' photons. All of this formed part of Horne's PhD thesis, accepted in 1970; but by then the experimental side was moving on.

In 1968, Shimony and Horne learned of experiments carried out at the University of California, Berkeley, by Carl Kocher and Gene Commins. They had measured the polarization of photons produced from calcium atoms in a process known as a cascade. They had taken measurements for just two polarizations at right angles to each other, in a simple experiment originally intended as a demonstration for an undergraduate physics course, and were unaware of Bell's work, so their results were inconclusive as far as testing hidden variables theory went; but clearly such an experimental setup could be adapted for testing Bell's theorem. Kocher and Commins weren't interested in pursuing this possibility, but Shimony and Horne now had a clear idea of what kind of experiment they wanted done. All they had to do was find a laboratory with the right kind of apparatus and an experimenter willing to do the job. They found the apparatus (actually using a mercury cascade rather than a calcium cascade) in the laboratory of Frank Pipkin, a professor at Harvard; and they found their experimenter in the form of Richard Holt, a graduate student at Harvard. With Pipkin's approval, Holt (who already knew of Bell's papers) set out on what turned out to be a mammoth project to measure Bell's inequality. But he had scarcely started when the team was hit by a bombshell, in the form of the programme for the spring 1969 meeting of the American Physical Society. Included in that bulletin was the abstract of a paper to be presented at the meeting by John Clauser, of Columbia University, who, the abstract revealed, was already working on the design of a similar experiment.

Clauser had graduated in physics from Caltech in 1964, then moved on to Columbia to work for an MA (1966) and PhD (1969). He was an experimenter, and his PhD work was in astrophysics – specifically, radio astronomy – but he was also interested in quantum physics, and had been astonished by Bell's papers, which he came across in 1967. His reaction was that the implications could not possibly be true, and he tried to find a counter-argument, but failed.[19] Realizing that the puzzle could only be definitively resolved by carrying out an experiment, he began scouring the scientific literature for papers describing such experiments. He also wrote to Bell, Bohm and de Broglie asking if they knew of any such experiments; the answers, of course, were all 'no'. Bell later said that this was the first communication he had ever received about the paper he had published in 1964.

Having failed to find any relevant experiments (although he did find the paper by Kocher and Commins), Clauser began to work out how he could do such an experiment himself, to the frustration of his thesis supervisor, who told him he was wasting his time and should concentrate on his radio astronomy work. From the outset, his objective was to find evidence that Bell's inequality was *not* violated, and that the world really did operate in accordance with local reality. So he prepared a paper describing the kind of experiment he planned to do, and arranged to present it at the spring 1969 meeting of the American Physical Society.

Clauser's abstract prompted Shimony to telephone him with the news that his team was thinking along the same lines, and a suggestion that they might join forces. Clauser was not keen on the idea, until he learned that Holt already had the basis of the experimental apparatus needed to do the job. The four scientists (Clauser, Horne, Shimony and Holt) met up at the American Physical Society gathering, and got on well enough together to produce a joint paper (usually referred to in the trade by their initials, CHSH). This paper presented a generalization of Bell's theorem and gave practical details of the kind of experiment needed to test it, using polarized photons. But before the collaboration could proceed further, Clauser was offered a job at Berkeley to work in radio astronomy under the laser pioneer and Nobel Prize winner Charles Townes. This was a plum position in itself, but Clauser was just as interested in the fact that the Kocher and Commins apparatus was still at Berkeley, and might be adapted to carry out a proper test of Bell's theorem. Commins was not keen on the idea, because he regarded entanglement as a known feature of quantum mechanics and saw no need to test it; but Townes, whose opinion carried more weight, was supportive, in spite of the fact that Clauser was supposed to be working on radio astronomy. The upshot was that Commins allowed a new graduate student, Stuart Freedman, to work on the project with Clauser, and Clauser never actually did any radio astronomy work worth mentioning. 'Without Townes,' says Clauser, 'I could never have done that experiment.' Clauser and Freedman were now competing with Holt, in a race to carry out the first proper test of Bell's theorem. But both teams had seriously underestimated the amount of time and effort this would take.

Fortunately, we do not have to go over all the trials and tribulations that the experimenters encountered. The bottom line is that in April 1972 Freedman and Clauser were able to publish a paper in the journal *Physical Review Letters* reporting that Bell's inequality was violated. 'We consider', they said, 'these results to be strong evidence against local hidden-variable theories.' Remember that this was the *opposite* of what Clauser had set out to prove; it is, somehow, more compelling when experimenters find their expectations wrong than when they find what they hoped to find. Freedman was awarded his PhD for his part in this work in May 1972.

Meanwhile, Holt had found the opposite result! His experiment implied, but not very strongly, that Bell's inequality was not violated. A wealth of other experiments have since confirmed that he was wrong and Clauser and Freedman were right, but nobody has ever found out exactly what went wrong with Holt's experiment; the most likely explanation is that a glass tube in the apparatus had an undetected curve in it which affected the polarization of the photons passing through it. Nevertheless, Holt received his

PhD in 1972. Clauser, in a heroic effort to resolve the confusion, carried out his own version of Holt's experiment and found results in disagreement with local realism. These results were published in 1976. The same year, another researcher, Ed Fry at Texas A&M University, had carried out a third test of Bell's theorem, using a laser-based system, and also found a violation of Bell's inequality.

In 1978, Clauser and Shimony published a review summarizing the situation, and concluding: 'It can now be asserted with reasonable confidence that either the thesis of realism or that of locality must be abandoned . . . The conclusions are philosophically startling: either one must totally abandon the realistic philosophy of most working scientists or dramatically revise our concepts of space-time.'

As these words make clear, by then it was almost impossible to find a loophole which would allow for the possibility of local reality. Almost, but not quite. One notable loophole remained; but it was about to be closed by an experiment carried out in Paris by a team headed by Alain Aspect.

Closing the loophole

The essence of the experiments to test Bell's theorem is that photons from a single source fly off in opposite directions, and their polarizations at various angles across the line of sight are measured at detectors as far away as possible from the source. The angle of polarization being measured can be chosen by setting one detector – a polarizing filter – at a particular angle (let's call it filter A), and another filter (filter B) at another carefully chosen angle on the other wing of the experiment. The number of photons passing through filter A can be compared with the number of photons passing through filter B. The results of the first-generation experiments, including those of John Clauser, showed that the setting of filter A affected the number of photons passing through filter B. Somehow, the photons arriving at B 'knew' the setting of A, and adjusted their behaviour accordingly. This is startling enough, but it does not yet prove that the communication between A and B is happening faster than light (non-locally), because the whole experimental setup is determined before the photons leave the source. Conceivably, some signal could be travelling between A and B at less than the speed of light, so that they are in some sense coordinated, before the photons reach them. This would still be pretty spooky, but it would not be non-local.

John Bell expressed this clearly, in a paper first published in 1981.[20] After commenting that 'those of us who are inspired by Einstein' would be happy to discover that quantum mechanics might be wrong, and that 'perhaps Nature is not as queer as quantum mechanics', he went on:

> But the experimental situation is not very encouraging from this point of view. It is true that practical experiments fall far short of the ideal, because of counter inefficiencies, or analyzer inefficiencies, [or other practical difficulties]. Although there is an escape route there, it is hard for me to believe that quantum mechanics works so nicely for inefficient practical set-ups and yet is going to fail badly when sufficient refinements are made. Of more importance, in my opinion, is the complete absence of the vital *time* factor in existing experiments. The analyzers are not rotated during the flight of the particles. Even if one is obliged to admit some long range influence, it need not

travel faster than light – and so would be much less indigestible. For me, then, it is of capital importance that Aspect is engaged in an experiment in which the time factor is introduced.

That experiment bore fruit soon after Bell highlighted its significance. But it had been a long time in the making.

Alain Aspect was born in 1947, which makes him the first significant person in this book to be younger than me (by just a year). He was brought up in the south-west of France, near Bordeaux, and had a childhood interest in physics, astronomy and science fiction. After completing high school, he studied at the École Normale Supérieure de Chachan, near Paris, and went on to the University of Orsay, completing his first post-graduate degree, roughly equivalent to an MPhil in the English-speaking world and sometimes known in France as the 'little doctorate', in 1971. Aspect then spent three years doing national service, working as a teacher in the former French colony of Cameroon. This gave him plenty of time to read and think, and most of his reading and thinking concerned quantum physics. The courses he had taken as a student in France had covered quantum mechanics from a mathematical perspective, concentrating on the equations rather than the fundamental physics, and scarcely discussing the conceptual foundations at all. But it was the physics that fascinated Aspect, and it was while in Cameroon that he read the EPR paper and realized that it contained a fundamental insight into the nature of the world. This highlights Aspect's approach – he always went back to the sources wherever possible, studying Schrödinger's, or Einstein's, or Bohm's original publications, not second-hand interpretations of what they had said. But it was only when he returned to

France, late in 1974, that he read Bell's paper on the implications of the EPR idea; it was, he has said, 'love at first sight'.[21] Eager to make a contribution, and disappointed to find that Clauser had already carried out a test of Bell's theorem, he resolved to tackle the locality loophole as the topic for his 'big doctorate'.

Under the French system at the time, this could be a large, long-term project provided he could find a supervisor and a base from which to work. Christian Imbert and the Institute of Physics at the University of Paris-South, located at Orsay, agreed to take him on, and as a first step he visited Bell in Geneva early in 1975 to discuss the idea. Bell was enthusiastic, but warned Aspect that it would be a long job, and if things went wrong his career could be blighted. In fact, it took four years to obtain funding and build the experiment and two more years to start to get meaningful results, and Aspect did not receive his big doctorate (*doctorat d'état*) until 1983. But it was worth it.

Such an epic achievement could not be attained alone, and Aspect led a team that included Philippe Grangier, Gérard Roger and Jean Dalibard. The key improvement over earlier tests of Bell's theorem was to find, and apply, a technique for switching the polarizing filters while the photons were in flight, so that there was no way relevant information could be conveyed between A and B at less than light speed. To do this, they didn't actually rotate the filters while the photons were flying through the apparatus, but switched rapidly between two different polarizers oriented at different angles, using an ingenious optical-acoustic liquid mirror.

In this apparatus, the photons set out on their way towards the polarizing filters in the usual way, but part of the way along their journey they encounter the liquid mirror. This is simply a tank of water, into which two beams of ultrasonic sound waves can be propagated. If the sound is turned off, the photons go straight through the water and arrive at a polarizing filter set at a certain angle. If the sound is turned on, the two acoustic beams interact to create a standing wave in the water, which deflects the photons towards a second polarizing filter set at a different angle. On the other side of the experiment, the second beam of photons is subjected to similar switching, and both beams are monitored; the polarization of large numbers of photons is automatically compared with the settings of the polarizers on the other side. It is relatively simple to envisage such an experiment, but immensely difficult to put it into practice, matching up the beams and polarizers, and recording all the data automatically – which is why the first results were not obtained until 1981, and more meaningful data not until 1982. But what matters is that the acoustic switching (carried out automatically, of course) occurred every 10 nanoseconds (1 ns is one billionth of a second), and it occurred *after* the photons had left their source. The time taken for light to get from one side of the experiment to the other (a distance of nearly 13 metres) was 40 ns. There is no way that a message could travel from A to B quickly enough to 'tell' the photons on one side of the apparatus what was happening to their partners on the other side of the apparatus, unless that information travelled faster than light. Aspect and his colleagues discovered that even under these conditions Bell's inequality is violated. Local realism is not a good description of how the Universe works.

Partly because there had been a groundswell of interest in Bell's work since the first pioneering Clauser experiment, and partly because of the way it closed the 'speed of light' loop-hole, Aspect's experiment made a much bigger splash than the first-generation experiments, and 1982 is regarded as the landmark year (almost 'year zero' as far as modern quantum theory is concerned) in which everything changed for quantum mechanics. One aspect of this change was to stimulate research along lines that led towards quantum computers. Another result of Aspect's work was that many other researchers developed ever more sophisticated experiments to test Bell's theorem ever more stringently; so far, it has passed every test. But although experiments like the ones carried out by Aspect and his colleagues are couched in terms of 'faster than light signalling', it is best not to think in terms of a message passing from A to B at all. What Bell's theorem tells us, and these experiments imply, is that there is an *instantaneous* connection between two (or more) quantum entities once they have interacted. The connection is not affected by distance (unlike, say, the force of gravity or the apparent brightness of a star); and it is specific to the entities that have interacted (only the photons in Aspect's experiment are correlated with one another; the rest of the photons in the Universe are not affected). The correlated 'particles' are in a real sense different aspects of a single entity, even if they appear from a human perspective to be far apart. That is what entanglement really means. It is what non-locality is all about.

It's worth reiterating that this result, non-locality, is a feature of the Universe, irrespective of what kind of theory of physics is used to describe the Universe. Bell, remember, devised his theorem to test quantum mechanics, in the hope of proving that quantum mechanics was not a good description of reality. Clauser, Aspect and others have shown that quantum mechanics *is* a good description of reality; but, far more important than that, they have shown that this is true because the Universe does not conform to local reality. Quantum physics is a good description of the Universe partly because quantum mechanics also does not conform to local reality. But *no* theory of physics that is a good description of the Universe can conform to local reality.

This is clearly Nobel-Prize-worthy stuff. Unfortunately, John Bell did not live long enough to receive the prize. On 30 September 1990, just a few days after receiving a clean bill of health at a regular checkup, he died unexpectedly of a cerebral haemorrhage. He was just sixty-two. Unbeknown to Bell, he had, in fact, been nominated for the Nobel Prize in physics that year, and although it usually takes several years of nominations for a name to rise to the top of the list, there is no doubt he would have got one sooner rather than later. The surprise is that Clauser and Aspect have not yet been jointly honoured in this way.

As is so often the case in quantum physics, there are several different ways of understanding how we get from Bell's theorem and entanglement to quantum computation. One way of getting an insight into what is going on – the one I prefer – is to invoke what is called the 'Many Worlds' (or 'Many Universes') Interpretation of quantum mechanics – in everyday language, the idea of parallel universes. Although he was not attracted by the idea, Bell admitted to Paul Davies that:

There is some merit in the many-universes interpretation, in tackling the problem of how something can apparently happen far away sooner than it could without faster-than-light signalling. If, in a sense, everything happens, all choices are realized (somewhere among all the parallel universes), and no selection is made between the possible results of the experiment until later (which is what one version of the many-universes hypothesis implies), then we get over this difficulty.[22]

Bizarre though it may seem, this is exactly the view espoused by the man who kick-started the modern study of quantum computation in the 1980s, David Deutsch. But that story belongs in Part Three of this book.

Second Interlude

Quantum Limits

When someone such as Richard Feynman says that the Universe is digital, it is the same as saying that it is 'quantized', as in quantum physics. Binary digits – bits – are quantized. They can either be 0 or 1, but they cannot be anything in between. In the quantum world, everything is digitized. For example, entities such as electrons have a property known as spin. This name is an unfortunate historical accident, and is essentially just a label; you should not think of the electron as spinning like a top. An electron can have spin ½ or spin −½ but it cannot have any other value. Electrons are part of a family of what we used to think of as particles, all of which have half-integer spin – ½, ³⁄₂ and so on. These are known as fermions. The other family of particles that make up the everyday world, known as bosons, all have integer spin – 1, 2 and so on. But there are no in-between values. A photon, a 'particle of light', is a boson with spin 1. This kind of quantization, or digitization, applies to everything that has ever been measured in the quantum world. So it is a natural step for quantum physicists to suppose that at some tiny scale, beyond anything that can yet be measured, space and time themselves are quantized.

The scale on which the quantization, or graininess, of space would become noticeable is known as the Planck length, in honour of Max Planck, the German physicist who, at the end of the nineteenth century, made the breakthrough which led to the realization that the behaviour of light could be explained in terms of photons. The size of the Planck length is worked out from the relative sizes of the constant of gravity, the speed of light, and a number known as Planck's constant, which appears at the heart of quantum mechanics – for example, the energy of a photon corresponding to a certain frequency (or colour) of light is equal to that frequency multiplied by Planck's constant. The Planck length is 0.0000000000000000000000000000000001 cm, or 10^{-33} cm in mathematical notation. A single proton is roughly 10^{20} Planck lengths across,[1] and it is no surprise that the effects of this graininess do not show up even in our most subtle experiments.

The smallest possible interval of time (the quantum of time) is simply the time it would take light to cross the Planck length, and is equal to 10^{-43} seconds. One intriguing consequence of this is that as there could not be any shorter time, or smaller time interval, then within the framework of the laws of physics as understood today we have to say that the Universe came into existence (was 'born', if you like) with an age of 10^{-43} seconds. This has profound implications for cosmology, but this is not the place to go into them.

It also has profound implications for universal quantum

simulators. The important point, which Feynman emphasized in his 1981 MIT lecture, is that if space itself is a kind of lattice and time jumps discontinuously, then everything that is going on in a certain volume of space for a certain amount of time can be described in terms of a finite number of quantities – a finite number of digits. That number might be enormous, but it is not infinite, and that is all that matters. *Everything* that happens in a finite volume of space and time could be exactly simulated with a finite number of logical operations in a quantum computer. The situation would be very similar to the way physicists analyse the behaviour of crystal lattices, and Feynman was able to show that the behaviour of bosons is amenable to this kind of analysis, although in 1981 he was not able to prove that all quantum interactions could be imitated by a simulator. His work was, however, extended at MIT in the 1990s by Seth Lloyd, who proved that quantum computers can in principle simulate the behaviour of more general quantum systems.

There's another way of thinking about the digitization of the world. Many accounts of the quantum world imply, or state specifically, that the 'wave' and 'particle' versions are of equal status. I've said so myself. But are they? It is a fundamental feature of a wave that it is continuous; it is a fundamental feature of a particle that it is not continuous. A wave, like the ripples spreading out from a pebble dropped in a still pond, can spread out farther and farther, getting weaker and weaker all the time until, far away from the place where the pebble was dropped, the ripples can no longer be detected at all. But either a particle is there or it isn't.

Light is often regarded as a wave, the ripples in something called an electromagnetic field. But those ripples, if they exist, do not behave like ripples in a pond. The most distant objects we can detect in the Universe are more than 10 billion light years away, and light from them has been travelling for more than 10 billion years on its way to us. It is astonishing that we can detect it at all. But what is it that we actually detect? *Not* a very faint ripple of a wave. Astronomers actually detect individual photons arriving at their instruments, sometimes literally one at a time. As Feynman put it, 'you put a counter out there and you find "clunk," and nothing happens for a while, "clunk," and nothing happens for a while'.[2] Each 'clunk' is a photon. Images of faint objects can be built up over many hours by combining these photons to make a picture – in one outstanding example, an image known as the Hubble Ultra-Deep Field was built up using photons gathered over nearly a million seconds (277 hours) of observation time on the Hubble Space Telescope. The fainter the object, the fewer photons come from it each second, or each million seconds; but each photon is the same as an equivalent photon from a bright object. Red photons always have a certain energy, blue photons a different energy, and so on. But you never see half, or any other fraction, of a photon; it's all or nothing. Which is why it is possible to simulate the Universe using a digital computer (provided it is a quantum computer). 'You don't find a tiny field, you don't have to imitate such a tiny field, because the world that you're trying to imitate, the physical world, is not the classical world, and it behaves differently,' said Feynman. 'All these things suggest that it's really true, somehow, that the physical world is representable in a discretized way.' This is the most important insight to carry forward into our discussion of quantum computers. Quantum computers actually provide a *better* representation of reality than is provided by our everyday experiences and 'common sense'.

PART THREE

Ion trap array incorporating a junction and linear ion trap sections, used by Winfried Hensinger and colleagues to demonstrate ion transport through a junction for the first time. Such a device is an important building block for a large-scale quantum computer.

CHAPTER FIVE

Deutsch and the Multiverse

The great appeal of the Many Worlds Interpretation of quantum mechanics (MWI) is that it avoids the problem of the collapse of the wave function, for the simple reason that the wave function never collapses. The problem with other interpretations – often known as the measurement problem – is deciding at what point between the quantum world and the everyday world the collapse occurs. This is what the Schrödinger's Cat puzzle is all about. Physicists have few qualms about accepting the possibility of a radioactive atom being in a superposition of states, but we all have qualms about the cat being in such a superposition. Does the collapse happen when the detector measures the radioactive material to see if it has decayed? Or is the cat's consciousness necessary to make the collapse happen? Could an ant be 'aware' enough to cause the collapse? Or a bacterium? These are not facetious questions, because larger and larger molecules have been sent through the experiment with two holes and behaved in line with quantum mechanics; there is even talk of doing it with molecules of DNA, if not yet with bacteria or cats. John Bell highlighted the ludicrousness of trying to apply the Copenhagen Interpretation to the Universe as a whole:

> Was the world wave function waiting to jump for thousands of millions of years until a single-celled living creature appeared? Or did it have to wait a little longer for some more highly qualified measurer – with a PhD? If the theory is to apply to anything but idealized laboratory operations, are we not obliged to admit that more or less 'measurement-like' processes are going on more or less all the time more or less everywhere?[1]

Everett sets the scene

The Many Worlds Interpretation avoids these difficulties by saying, for example, that in the case of the cat in the box there are two universes, one with a dead cat and one with a live cat; and similarly in other situations, every quantum possibility is realized. In the mid-1950s the American researcher Hugh Everett III put the MWI on a proper mathematical footing, and showed that for all practical purposes it is exactly equivalent to the Copenhagen Interpretation. Since this meant it made no difference to their calculations, most practising quantum mechanics ignored it. Unfortunately, there was a flaw in Everett's presentation of the idea which meant that even the few theorists who did think about the implications also found it hard to take seriously.

Everett described the many worlds of his model in terms of splitting. In the case of Schrödinger's cat, this would mean that in the course of the 'experiment' we start out with a single cat in a single universe (or world) and that the world then splits into two, one with a live cat and one with a dead cat. Everett used a different analogy, at least in the first draft of his idea, which he showed to his supervisor at Princeton, John Wheeler, in the autumn of 1955. He used the word 'splitting', and made an analogy with the splitting of an amoeba. You start with one amoeba, and then have two, each of which, if it had a memory, would remember the same experiences, or history, up to the point of the split. At that point, the two individuals go their separate ways, eventually to split again, and their offspring split in their turn, and so on. Wheeler 'persuaded' Everett to leave the amoeba analogy out of his PhD thesis and the published version of his work, which appeared in the journal *Reviews of Modern Physics* in 1957. But Everett did say in print that 'no observer will ever be aware of any "splitting" process', and clearly had the idea of one 'history' branching repeatedly as time progressed.[2] This was spelled out by Bryce DeWitt, an enthusiastic supporter of Everett's idea, who wrote: 'Every quantum transition taking place in every star, in every galaxy, in every remote corner of the universe is splitting our local world on Earth into myriad copies of itself.' Or, as John Bell put it: 'Quite generally, whenever there is doubt about what can happen, because of quantum uncertainty, the world multiplies so that all possibilities are actually realized. Persons of course multiply with the world, and those in any particular branch would experience only what happens in that branch.'[3]

Again, the language is that of branching and multiplication of worlds by splitting. Bell is not enthusiastic, but, almost in spite of himself, does not dismiss the idea out of hand:

The 'many worlds interpretation' seems to me an extravagant, and above all an extravagantly vague, hypothesis. I could almost dismiss it as silly. And yet . . . It may have something distinctive to say in connection with the 'Einstein Podolsky Rosen puzzle', and it would be worthwhile, I think, to formulate some precise version of it to see if this is really so. And the existence of all possible worlds may make us more comfortable about the existence of our own world . . . which seems to be in some ways a highly improbable one.[4]

Although Wheeler was initially enthusiastic about Everett's idea, as the years passed he developed qualms. Two decades later, he said: 'I confess that I have reluctantly had to give up my support of that point of view in the end – much as I advocated it in the beginning – because I am afraid it carries too great a load of metaphysical baggage.'[5] It seems to me to carry a lesser load of metaphysical baggage than the idea of the collapse of the wave function, even in this imperfect form. But there is a more serious objection to Everett's version of MWI than metaphysics.

For the Everett MWI seems to contain the same flaw, the measurement problem, as the Copenhagen Interpretation itself. It's just that in one case the measurement puzzle refers to the moment of collapse, and in the other it refers to the moment of splitting. This might look like the death knell for the idea, at least as an alternative to the Copenhagen Interpretation; but Everett (and Wheeler, DeWitt and other supporters of the idea) missed a trick.

I confess that I missed the same trick, long ago, although I had fewer qualms than Bell about espousing the MWI, which I enthusiastically endorsed in my book *In Search of*

Schrödinger's Cat. But someone who did not miss this trick was Schrödinger himself, as I learned to my surprise when writing his biography.

Solving the measurement problem

In 1952, Schrödinger published a scientific paper entitled 'Are There Quantum Jumps?' Arguing that there is no reason for a quantum superposition to collapse just because we look at it, or because it is measured, he said that 'it is patently absurd to let the wave function be controlled in two entirely different ways, at times by the wave equation, but occasionally by direct interference of the observer, not controlled by the wave equation'. His solution was that the wave function does not collapse, and no choice is ever made between a superposition of states. Although Schrödinger himself – perhaps surprisingly – never pointed out the implications in terms of his famous cat puzzle, this neatly demonstrates the point he is making: in effect, he is saying that in the cat experiment, the wave functions leading to the 'dead cat' and the 'live cat' are equally real, and remain so both before and after the box is opened. In everyday language, there are two parallel worlds, one with a live cat and one with a dead cat; *and* – this is the crucial point – there *always were* two worlds, each starting out with a live cat but becoming different when one of the cats, but not the other, dies. There is no splitting, and no measurement problem. For once, the popular term 'parallel worlds' is the most apt, and removes the image of branching realities from our minds. The two worlds (or universes) have identical histories up until the point where the experiment is carried out, but in one universe the cat lives and in the other the cat dies. They are like parallel lines, running alongside each other. And running on either side of those two parallel worlds are more parallel worlds, each slightly different from its immediate neighbours, with close neighbours having very similar histories and widely separated universes differing more significantly from one another. This is not, strictly speaking, an 'interpretation' at all; it is, as Schrödinger pointed out, what the equations tell us. It is the simplest way to understand those equations. If we ever did an experiment like the one Schrödinger envisaged, we would not be forcing the universe to split into multiple copies of itself, but merely finding out which reality we inhabit.

The response of the few people who noticed this idea at the time was summed up rhetorically by Schrödinger himself in a talk he gave in Dublin in 1952. I have quoted it before, but it is surely worth quoting again:

Nearly every result [the quantum theorist] pronounces is about the probability of this or that or that . . . happening – with usually a great many alternatives. The idea that they may not be alternatives but all really happen simultaneously seems lunatic to him, just impossible. He thinks that if the laws of nature took this form for, let me say, a quarter of an hour, we should find our surroundings rapidly turning into a quagmire, or sort of a featureless jelly or plasma, all contours becoming blurred, we ourselves probably becoming jelly fish. It is strange that he should believe this. For I understand he grants that unobserved nature does behave this way – namely according to the wave equation. The aforesaid alternatives come into play only when we make an observation – which need, of course, not be a scientific observation. Still it would seem that, according to the quantum theorist, nature is prevented from rapid jellification only by our perceiving or observing it . . . it is a strange decision.

All of this has led me to change my view on the nature of quantum reality. As Bell's theorem and the experiments described in Chapter 4 make clear, the world is either real but non-local, or local, but not real. In *In Search of Schrödinger's Cat*, I came down in favour of locality, and concluded (echoing John Lennon) that 'nothing is real', at least until it is measured. Now, I am inclined to accept non-locality, with the corollary that the world is real – or rather, that the many worlds are *all* real. Not 'nothing is real', but 'everything is real', since wave functions never collapse.[6]

If Everett had been aware of Schrödinger's views when he came up with his own version of MWI a few years later, he could have produced an even more satisfactory package of ideas than he did. But it is still unlikely that many people would have taken it very seriously until the experimental proof of Bell's theorem had arrived. Schrödinger's insight languished in obscurity for more than thirty years, and when David Deutsch elaborated the modern version of the Many Worlds Interpretation in the 1980s he did so without drawing directly on this aspect of Schrödinger's work – indeed, without being aware of Schrödinger's contribution to MWI.

The worlds of Deutsch

David Deutsch is an unusual physicist with unconventional habits. Although he doesn't like too much attention being given to his eccentricities, since he feels this may divert attention from the underlying importance of his work, and might be taken as implying (incorrectly) that you have to be weird to be creative, the stories are as irresistible (if as irrelevant) as those about Einstein not wearing socks or Turing's bicycle chain. Deutsch lives in an ordinary, rather unkempt-looking house in a suburb of Oxford, and as I discovered the visitor has to be prepared to negotiate their way from the front door past piles of boxes and papers to the darkened room, dominated by computer screens, in which he works. The curtains are almost invariably drawn shut, and Deutsch tends to work at night and sleep (a little) in the day – lunchtime is typically around 8 pm, followed by a solid twelve hours' work. Although he is affiliated with the Centre for Quantum Computation at Oxford's Clarendon Laboratory, and is a non-stipendiary visiting professor of physics, Deutsch has no paid academic post, living off lecturing and writing (plus the proceeds of various prizes he has been awarded[7]); colleagues are more likely to encounter him at an international meeting in some far-off land than among the dreaming spires of Oxford.

A (true) story that has become part of Oxford folklore tells how a Japanese film crew who came to interview Deutsch were so concerned by the mess that they offered to tidy it up. Deutsch explained that what looked like a mess to them was order to him, and he knew what was in each pile of papers; but he reluctantly agreed to their request, on condition that they promised to put everything back afterwards as they had found it. So they photographed everything, made notes like records of an archaeological dig, cleared away what they regarded as a mess – and, after the interview, restored everything to its original place, so that Deutsch could instantly locate anything in his 'filing system'. I can empathize with his insistence, since I operate a similar system myself.

So how did Deutsch get into this position? He was born in 1953 in Haifa, Israel. After studying at William Ellis School in London, he took his first degree (in natural sciences) at the University of Cambridge, as a member of Clare College, staying on for the Part III in Mathematics (the Cambridge equivalent of a Master's degree), awarded in 1975. Deutsch then moved to Oxford, studying for his doctorate in mathematical physics under the supervision of Dennis Sciama. These studies also took him to the University of Texas at Austin, where he was supervised by John Wheeler – Everett's PhD supervisor who subsequently had doubts about the MWI – and stayed on to do post-doctoral research.

Always interested in fundamental physics, Deutsch had chosen for his thesis topic to investigate the theory of quantum fields in curved space–time, a problem that involves both quantum theory and the general theory of relativity, the two most fundamental theories of the world that we have. Unfortunately, at first sight the two theories seem to be incompatible with one another. The Holy Grail of physics – on which no one has yet laid hand – is the uniting of these two great theories in one package, quantum gravity. Deutsch hoped that understanding quantum fields in curved space–time would provide a clue to quantum gravity; that hope was not fulfilled, but in the course of his investigation he decided that quantum theory contained deeper truths about the nature of reality than the general theory of relativity did, and decided to focus on quantum theory in future.

While in Texas, Deutsch also worked under the super-vision of Bryce DeWitt, who was almost single-handedly responsible for reviving Everett's MWI, largely ignored since its publication in his 1957 paper. From DeWitt Deutsch learned about MWI, and in 1977 he was in the audience when Everett gave a four-hour tour de force presentation of his ideas at a conference in Austin organized by DeWitt. This was the only major exposition of his ideas ever given by Everett, who had moved straight from his PhD studies in Princeton (before his MWI paper was even published) out of academia into the secret world of the Pentagon, working in the Weapons Systems Evaluation Group, at the height of the Cold War. Deutsch and Everett discussed the many universes idea over lunch, and Deutsch became convinced that this was the right way to understand quantum mechanics.

Almost immediately, Deutsch came up with the idea of a self-aware machine that could test the many worlds hypothesis. The idea had been thought untestable, and therefore arguably not 'real science', but by 1978 he had devised a thought experiment involving a machine which would be aware of the existence of more than one reality producing interference. Only later did he appreciate that this would, in fact, be a version of a quantum computer. Deutsch described this experiment in an interview with Paul Davies broadcast by the BBC in 1982.[8] By then, he had returned to Oxford (in 1979), where he has been based ever since.

The experiment Deutsch described to Paul Davies 'requires the observation of interference effects between two different states of an observer's memory'. The 'observer' would be a machine operating on quantum principles – a quantum computer, although he did not use the term at that time. 'The experiment hinges on observing an interference phenomenon inside the mind of this artificial observer . . . by his trying to remember various things so that he can conduct an experiment on his own brain while it's working.' Deutsch refers to a 'quantum memory unit' which observes a feature of the state of an atomic system, such as its spin. The experiment can easily be set up so that the system is in a superposition, before interference occurs. In one of the parallel universes the brain will be aware that the atomic spin is up; in the other it will be aware that the atomic spin is down. But it will *not* observe both at once. At this stage, it can communicate with outside observers to confirm to its human colleagues that it is experiencing one possibility, and only one (in other words, it is *not* experiencing a superposition) – but it doesn't tell them which one, because interference can only take place in systems that are not entangled with the rest of the universe. Then, interference takes place, and in *both* universes the result is the same: equivalent, if Everett is correct, to the interference produced by electrons going through both holes of the double slit experiment. The two universes have become identical to one another (at least in this respect), but although each of them records interference, each also contains proof that there was only one history in that universe:

> If interference occurs he [the quantum computer] can infer that these two possibilities must have existed in parallel in the past – supporting the Everett interpretation. However, if the conventional interpretation is true,[9] [the wave function will collapse and] although it'll still be true that he will write down 'I am observing only one', by the time he gets to the interference phenomenon it won't work (i.e. the interference won't occur). And so he will have demonstrated that the Everett interpretation is false.

This is not easy to get your head around, but in his book *The Beginning of Infinity* Deutsch offers a simpler proof of the reality of parallel worlds. It's all based on the properties of half-silvered mirrors.

A half-silvered mirror is the basis of the 'one-way' mirror,

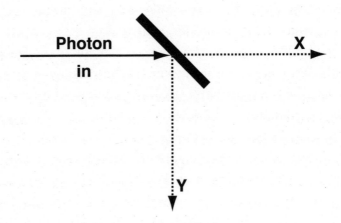

Semi-silvered mirror

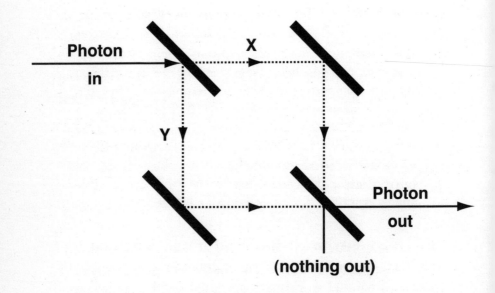

Mach–Zehnder interferometer

which looks like a mirror from one side but can be seen through like a sheet of plain glass from the other side. It works because half of the photons that hit the mirror are reflected, and half are transmitted through it. By tilting the mirror at 45 degrees to a beam of light, it can be arranged for the reflected photons to be deflected at a right-angle, while the transmitted photons go straight on, as in the top diagram. If photons are fired one at a time towards the mirror, half of them are transmitted along path X and half deflected along path Y, but you cannot have half a photon, so no individual photon divides to follow both paths. There is, of course, no way to tell in advance which path an individual photon will follow; there is a 50:50 chance of its following either path. But if you fire a million photons through the experiment you can be sure that almost exactly half a million will go each way.

Now comes the interesting part. Four mirrors can be set up in a square arrangement, called a Mach–Zehnder interferometer, as in the bottom figure. The mirrors at top left and bottom right of this diagram are half-silvered; the other two are conventional mirrors that reflect all the photons that strike them. If the top left mirror is removed and photons are sent along path X towards the top right mirror, they are reflected downwards and emerge from the bottom right mirror (half-silvered, remember) either to the right or downwards with 50:50 probability. In the same way, without the top left mirror, a photon sent down along path Y will arrive at the bottom right mirror and emerge either to the right or downwards with 50:50 probability. But with all the mirrors in place, as in the diagram, *all* the photons emerge travelling to the right, even when they are fired in one at a time. This can be explained in terms of the interference between path X and path Y, although I shall not go into the details here.[10] But since the effect is observed for single photons, which cannot be divided, how can both paths affect the outcome of the experiment? Deutsch argues persuasively that it is because of interference between parallel universes, in some of which the photon(s) take path X and in some of which the photon(s) take path Y. 'The fact that the intermediate histories *X* and *Y* *both* contribute to the deterministic final outcome makes it inescapable that both are happening at the intermediate time.' In other words, 'the multiverse is no longer perfectly partitioned into histories' during such an experiment. 'Quantum interference phenomena constitute our main evidence for the existence of the multiverse and what its laws are.'

In 1985, Deutsch published a scientific paper that is now regarded as marking the beginning of the quest for a quantum computer.[11] The inspiration for the paper came from a conversation with Charles Bennett of IBM, who opened Deutsch's eyes to the physical significance of the 'Church–Turing principle' – in Turing's words, that 'it is possible to invent a single machine which can be used to compute any computable sequence'. In the mid-1930s, Turing's insight pointed the way towards universal classical computers; in the mid-1980s, Deutsch's insight pointed the way towards universal quantum computers. He described a quantum generalization of the Turing machine, and showed that a 'universal quantum computer' could be built in accordance with the principle that 'every finitely realizable physical system can be perfectly simulated by a universal model computing machine operating by finite means . . . Computing machines resembling the universal quantum computer could, in principle, be built and would have many remarkable properties not reproducible by any [classical] Turing machine', although they would also be able to simulate perfectly any such Turing machine. He stressed that a strong motivation for developing such machines is that 'classical physics is false'. And in particular he drew attention to the way in which 'quantum parallelism' would allow such computers to perform certain tasks faster than any classical computer.

This parallelism is the reason for Deutsch's interest in quantum computing. He is not interested in the practicalities of building such computers or the things they might achieve, except as evidence for the existence of the Multiverse. In his landmark 1985 paper he stated baldly that 'quantum theory is a theory of parallel interfering universes', and posed an early version of a question to which he would repeatedly return in his writings: when a quantum computer carries out a calculation that requires two 'processor days' of computation in less than a day, *where was it computed?*[12] Now that we actually have quantum computers, the question assumes even greater significance; but most of the people who work with such machines ignore it, happy that their computers work FAPP. But the question cannot be swept under the rug, and before going on to the practicalities of quantum computing in the twenty-first century I shall summarize thinking about the quantum multiverse today.

A measure of universes

It's worth repeating that the important point to take away from all this is that even within the modern version of the Many Worlds Interpretation, interference, such as in the experiment with two holes, does not involve any 'splitting' and 'recombining' of the Universe(s). Rather, this kind of experiment allows interference between nearby parallel worlds to become obvious. 'Nearby' in this sense means 'similar'. Another important point is that it is meaningful to talk about fractions of infinity. Although half of infinity is still infinity, it does make mathematical sense to talk about 50:50 probabilities (as in the experiment with two holes, or a photon hitting a half-silvered mirror) in terms of half of the universes following one history and half following the alternative history.

Mathematicians study infinity in terms of infinite sets. In this context, an infinite set has the property that some part of the set has as many components (elements) as the whole set. The so-called natural numbers, or integers, provide an example. Counting 1, 2, 3 and so on, there is always one number bigger than the last number counted. But if we write numbers down in two columns alongside each other, starting with 1 in one column and 2 (or any other number bigger than 1) in the other, we get pairs, 1–2, 2–3, 3–4 and so on. Clearly, the second column has as many elements as the first column – there is a one-to-one correspondence. But the second set is also a subset of the whole set, because it does not contain the number 1. So the whole set is infinite.

Any infinity that can be placed in a one-to-one corres-pondence with the set of natural numbers is called a countable infinity. The set of 'all the natural numbers except 1' is itself an example of a countable infinity. There are other kinds of infinity, which are larger than countable infinities and are called uncountable infinities. An example is the set of all possible decimal numbers (known to mathematicians as 'real' numbers). We can see this by assigning decimal numbers in the range from 0 to 1 to the integers, so that, for example, 1 is paired with 0.12345 ..., 2 is paired with 0.23456 ..., 3 is paired with 0.34567 . . . and so on.[13] Now, make a new decimal number by making the first digit anything *except* the first digit of the first number (that is, not 1), the second digit anything *except* the second digit of the second number (not 3), the third digit anything *except* the third digit of the third number (not 5) and so on. This new decimal number will differ from the decimal number assigned to the natural number 1 in the first digit, from the number assigned to 2 in the second digit, from the number assigned to 3 in the third digit, and so on. So it does not correspond to any of the numbers that are paired with the natural numbers. And this is true for an uncountably large infinity of real numbers.

Unfortunately, as far as we can tell the Multiverse consists of an uncountably infinite set of universes. That includes, as I have discussed in my book *In Search of the Multiverse*, such exotic possibilities as universes in which the laws of physics are different from those in our Universe. But, happily, if we restrict ourselves to the set of all universes in which the laws of physics are the same as in our Universe (itself an uncountably large infinite set) it is meaningful to talk in terms of, say, 'half the universes' doing one thing and 'half the universes' doing something else, even though each half is itself infinite. This is because the phenomenon of quantum interference involves interactions between universes – internal interactions *within* the Multiverse, making the Multiverse itself a single entity – that make such talk in terms of proportions or ratios meaningful. This way of measuring infinities is known, logically enough, as a 'measure'. Deutsch has described the Multiverse as the set of all such universes evolving together, like a machine in which cogwheels are interconnected so that you cannot move one without moving the rest.

So when we talk about the cat in the box, we should not be thinking of a single cat in a single box, or even two cats in two boxes in two parallel worlds, but an uncountably infinite number of cats in an uncountably infinite number of boxes. In half of those parallel universes the cat dies, and in the other half it lives. For more subtle situations where there are different possible outcomes with different probabilities, it is meaningful to say that, for example, in 25 per cent of the universes one history develops, in 60 per cent an alternative history develops, and in 15 per cent a third history develops. And that is why, in situations such as radioactive decay, in any one universe the outcome is entirely random (that is, it obeys the laws of probability) from the point of view of an observer in that universe.

There's another aspect of the infinite Multiverse that is worth mentioning, even though it is not directly relevant to the story of quantum computation. In Deutsch's words, 'all fiction that does not violate the laws of physics is fact'. So all of Jane Austen's stories recount real events in parallel realities to our own; but *The Lord of the Rings* does not. And in an infinite number of universes there are writers busily producing what they regard as a fictional tale about a quantum physicist called David Deutsch who wrote a land-mark paper about the concept of a universal quantum computer.

The connection between the Multiverse and the quantum computer is made explicit by Deutsch using a concept known as 'fungibility', taken from the lexicon of legal language. In law, two objects are deemed to be fungible if they are identical FAPP. The usual example is a banknote. If I borrow a £10 note from you and promise to pay it back on Friday, you do not expect me to give you back the same £10 note; any £10 note will do. All (genuine) £10 notes are fungible in law. But if I borrow your car for the afternoon to visit my cousin, you expect me to give you the same car back in the evening; cars are not fungible. Elaborating the financial theme slightly, my wife and I have a joint bank account. Half the money is mine, and half is hers. But there is no sense in which there are two piles of money in the bank, labelled 'his' and 'hers' – and there wouldn't be even if the money was literally 'in the bank' and not merely an electronic record inside a computer. Deutsch argues that universes with identical histories are literally and physically fungible. This goes beyond the concept of parallel universes – which implies running side by side, somehow separated in space, or superspace, or whatever you want to call it. Two or more fungible universes become differentiated when forced to by quantum phenomena, such as the experiment with two holes, leading to the diversity of universes within the Multiverse, but in accordance with the laws of probability, as discussed above. These universes can briefly interfere with one another as this process occurs. But quantum interference is suppressed by entanglement, so the larger and more complex an object is, the less it will be affected by interference. Which is why we have to set up careful experiments (like the one with half-silvered mirrors) to see interference at work, and why there are such large practical problems involved in building quantum computers, where entanglement has to be avoided except in special places. A universe like ours, suggests Deutsch, is really made up of a bundle of 'coarse-grained' histories differing from each other only in sub-microscopic detail, but affecting each other through interference. Each such bundle of coarse-grained histories 'somewhat resembles the universe of classical physics'. And these coarse-grained universes really do fit the description of parallel universes familiar from science fiction.

Which brings us back to the quantum computer, where, as I explained in the Introduction, processing involves qubits, not bits, of information. The key feature of a qubit is that it can exist not only in the state 0 or the state 1, but in a superposition of both states. But they have another important property. Qubits, like other quantum entities, can be entangled with one another. In carrying out a calculation using a quantum computer, the set of instructions – the program, or algorithm – initially sets out the 'problem' by putting some of an array of qubits (the input register) into a superposition of states. If each qubit is regarded as a register for a number then it stores two (or more) numbers simultaneously. During the next stage, a computation is carried out which spreads information to other qubits in the array (the output register), but not into the outside world (stopping the information leaking out too soon and being destroyed by entanglement is one of the key practical problems involved in building quantum computers). Crudely speaking, each component of the input register is entangled with a corresponding component of the output register, like the photons in the EPR experiment. The information has been processed simultaneously in all the histories represented by the superposition of states – in all the parallel universes, in everyday language. Finally, the qubits are allowed to interfere in a controlled fashion that provides a combination of the information from all those histories – an 'answer'. The computer has carried out separate parts of the calculation in separate histories (separate universes) and produced an answer based on the interference between those histories.

But quantum computers are not magic. Like all tools, they have their limitations. There are some kinds of problem that they can solve with ease that classical computers cannot begin to tackle on any timescale relevant to human activities; but there are other problems that are not susceptible to this approach. When quantum computers are good, they are very, very good; but when they are out of their depth, they are not so much bad as no better than classical computers. To explain this, I'll start with the good.

The Good: cracking codes conveniently

The best example of the power of quantum computation is the one that has inspired (if that is the right word) large amounts of money to be spent on making it a practical reality – codebreaking. There is a pleasing resonance with the reason why so much effort was devoted to building Colossus, the first computer, but this time the impetus comes as much from big business as from the military.

Cryptography has, of course, moved on a long way since the days of Colossus and Enigma. Modern codes are devised specifically to be difficult, if not impossible, to break using classical computers. The classic example of an unbreakable code is the 'one time pad' system, where two protagonists (usually known as Alice and Bob) each have a pad made up of sheets of paper each listing a coding system – the first sheet might begin A codes as D, B codes as A, C codes as Z, and so on.[14] Alice encodes her message using the first sheet on the pad, then rips it out and destroys it. Bob receives the message and decodes it using the first sheet on his pad, then rips it out and destroys it. The next message uses sheet two of the code pad, and so on. The 'one time' aspect is vital, because if the same code is used more than once an eavesdropper (Eve) can crack the code using the kind of techniques pioneered at Bletchley Park. This method is utterly secure as long as Eve doesn't get hold of a copy of the pad, but suffers from the difficulty of getting fresh pads to Alice and Bob as required. This used to involve trusted couriers carrying locked brief-cases between embassies or the offices of global businesses in different parts of the world (and Enigma was essentially based on this kind of idea, generating its own 'pads' as it went along). But the system has been superseded by so-called public key cryptography.

This technique uses a freely available system – the public key – to encode messages, which can only be decoded by the intended recipient using a private key unknown to anyone else, including the sender of the message. Anybody can encrypt messages using Alice's public key, but only Bob can decrypt them. The mathematics behind this technique was first published in 1974, by Ralph Merkle, a graduate student at the University of California, Berkeley, and developed by Whitfield Diffie and Martin Hellman at Stanford University. The system depends on generating a public key using a combination of two numbers (M and N) which are openly published and a third 'secret' number (actually a third and a fourth number, since each cryptographer – sender and receiver – also has their own secret number). It is very difficult (though not actually impossible) to work out what the secret numbers are from M, N and the public key. The system was developed further in 1977 by a team at MIT: Ronald Rivest, Adi Shamir and Len Adleman. In this form, the cryptographic technique is known as the RSA algorithm, from their initials. Using RSA, all Alice has to do is publish a public key, which anyone can use to send her a message that she alone can read.

The names of all of the people mentioned in the previous paragraph are enshrined in the history of cryptography. It was only in 1997 that the British government released previously secret records revealing that the whole package of ideas had been developed earlier by mathematicians working at GCHQ, the heirs to the Bletchley Park codebreakers, and kept under wraps for obvious reasons.

Without delving deeply into the full mathematical details, the secrecy of codes based on the RSA algorithm depends on the difficulty of finding the factors of a very large number. Factors are something we all learned in school – the factors of the number 15, for example, are 3 and 5, because they are both primes and $3 \times 5 = 15$. Numbers can have more than two factors – $3 \times 5 \times 2 = 30$ – but in the present context we are interested only in numbers which are obtained by multiplying two large prime numbers together. These numbers are the factors of the resulting very large number. It's easy to find the factors of 15 by trial and error, but the difficulty grows exponentially as the size of the number increases. The reason – the only reason – Eve cannot crack Alice's code is because she does not know the factors of the large number used in Alice's public key. A key of 1,024 bits (one kilobit) would represent a decimal number with more than 300 digits; while such a number is simple to work with for coding purposes, finding its factors using a classical computer would take longer than the time that has elapsed since the Big Bang in which the Universe as we know it was born.[15] No wonder big business and the military make use of the RSA algorithm – which is also the basis (using much shorter keys) of the secure systems used to preserve your privacy when you use your credit card over the World Wide Web. So imagine the mixture of horror and delight when it was realized that even a very simple quantum computer could factorize this kind of 300-digit number – thereby cracking the code – in about four minutes. More advanced machines could do it in milliseconds. Horror, because 'our' secrets might be revealed; delight because 'their' secrets might be revealed to us.

If a quantum computer is asked the question 'What are the factors of 15?', after performing the calculation its output register contains information which can be processed using classical techniques to give answers corresponding to 3 and 5. When the output is read, it will give one of these numbers – and if you know one, you know the other. Even with large numbers, by repeating the process it is easy to find all the factors. But how does it perform the calculation?

The breakthrough came from Peter Shor, of Bell Labs, in 1994. In his landmark paper of 1985, David Deutsch himself had given an example of a kind of problem susceptible to parallel processing that could be solved twice as fast by a quantum computer using an algorithm he devised (the Deutsch algorithm) as by a classical computer; but this was not impressive enough to encourage efforts to build a working quantum computer, since it would be easier and cheaper to build two classical computers and use them working in parallel to solve such problems twice as fast. Deutsch himself was only interested in this example as proof of the reality of parallel universes. But the Shor algorithm was a whole new ball game.

The essence of Shor's algorithm is that long strings of numbers contain periodicities – patterns of repetition – which may not be obvious at first sight but can be teased out by mathematical analysis. The details are fascinating, and if you want them the best place to look is in Julian Brown's book *Minds, Machines and the Multiverse*, but what matters here is that periodic patterns can be probed using a technique known as Fourier analysis, which unpacks complicated wave patterns into their component parts. For example, the complicated sound wave corresponding to the sound of a group of musical instruments playing a chord together could be unravelled by Fourier analysis into a wave corresponding to a violin note, a wave corresponding to the cello, a wave corresponding to a clarinet and so on. Astronomers use the same sort of technique to do such things as, for example, analyse the variations in the output of light from a star and find underlying patterns of periodic variation which reveal details of the structure of the star.[16] Using a classical computer, it would be possible in principle to search for periodicities by considering each possibility in turn – looking for a pattern that repeats every two digits, then for one that repeats every three digits, then for one that repeats every four digits, and so on. But for numbers containing hundreds of digits this could take longer than the age of the Universe. A quantum computer using Shor's algorithm could set up a superposition with all the possible periods (waves) being examined simultaneously, and select the one that actually fitted the number in question – which would take only a few minutes. The periodicity found in this way could then be used by a classical computer to reveal the factors of that number.

One particularly attractive feature of this procedure is that you know at once if you have got the right answer – you just multiply the factors together to see if they give you the right number. If there has been a glitch inside the computer and you got the wrong answer (and the first quantum computers are likely to be at least as temperamental as the first classical computers were), you just run the problem again. Even if the quantum computer only works properly one time in a hundred, you can still solve the problem in a few hundred minutes: nothing compared with the age of the Universe.

Deutsch explains this in terms of a myriad identical computers in a myriad almost identical universes, each of which tests one of the possible periodicities, which then interfere so that each of them outputs the one correct answer. You might ask why the myriad codebreakers in all those universes bother to run their computers for our benefit. But FAPP those codebreakers are us; they also want to crack the code!

All of this provides Deutsch with his most powerful argument for the reality of the Multiverse. Bear in mind that a single 250-qubit 'memory' contains more bits of information than there are atoms in the observed Universe. As Deutsch writes in *The Fabric of Reality*:

> To those who still cling to a single universe world-view, I issue this challenge: *explain how Shor's algorithm works*.[17] I do not merely mean predict that it will work, which is simply a matter of solving a few uncontroversial equations. I mean provide an explanation. When Shor's algorithm has factorized a number, using 10^{500} or so times the computational resources that can be seen to be present, where was the number factorized? There are only about 10^{80} atoms in the entire visible universe, an utterly minuscule number compared with 10^{500}. So if the visible universe were the extent of physical reality, physical reality would not even remotely contain the resources required to factorize such a large number. Who did factorize it then? How, and where, was the computation performed?

'I have never', Deutsch told me, 'received a satisfactory answer to this question.' Or, as his colleague Artur Ekert, also of Oxford University, put it at that time: 'The Many Worlds Interpretation of quantum mechanics is the least weird of all the weird interpretations of quantum mechanics.' These days, however, Ekert has told me, 'I do not refer to the Many Worlds as an interpretation, for it is just quantum theory in its bare essentials.'[18]

On its own, the codebreaking potential of the Shor algorithm would justify the effort now going into attempts to build quantum computers. It isn't just future codes that will be broken. You can be sure that government agencies, doubtless including GCHQ, are already stockpiling intercepted coded messages they suspect of being of long-term political significance in the hope and expectation of being able to read them in ten or twenty years' time.

There is another extremely useful algorithm that will enable quantum computers to perform rapidly a class of tasks that would take an interminably long time on classical computers. This is the seemingly mundane job of searching through large numbers of pieces of information to find the one that is required. The usual example is the problem of searching a printed telephone directory to find the name of somebody when you only have their telephone number. Starting at either end of the alphabet, you have to check each number against the corresponding name; since the name could be anywhere in the list, on average if N is the number of numbers in the list it will take $N/2$ steps to find it. In 1996 Lov Grover, another Bell Labs researcher, found an algorithm which reduces this number to roughly the square root of N, by starting with a superposition of all the pieces of information and carrying out a series of operations which magnify (in a quantum sense) the one we are looking for. This is impressive, but not as impressive as Shor's algorithm, and not sufficient to displace classical computers as long as they remain cheap. The real importance of the discovery lay in providing evidence that quantum computing could be used for more than one thing, encouraging further investigation.

Remember, though, that 'classical physics is false'. Perhaps the most important, and certainly the most profound, feature of a quantum computer is that it would be able to simulate physics *precisely*, not just approximately. No matter how good the approximations involved in simulating the Universe in a classical computer may be, they can never be perfect because the Universe is not classical. And remember how Deutsch got interested in quantum physics originally – through the search for a quantum theory of gravity. If we are ever to have a satisfactory 'theory of everything' incorporating both quantum theory and gravity, it is almost certain that it will only be found with the aid of quantum computers to simulate the behaviour of the Universe.[19]

But if that is the long-term dream, we also have to face a harsher short-term reality. When they are good, quantum computers are very, very good. But . . .

The Bad: limits of quantum computation

Ironically, Grover's algorithm, which stimulated interest in quantum computation in the late 1990s, also provides a neat example of the limitations of quantum computing. In 1997, the year after Grover published his discovery, the computer Deep Blue made headlines by defeating the human world chess champion, Gary Kasparov. This encouraged speculation about how good a chess-playing quantum computer might be, if it were able to use Grover's algorithm to search for the best moves.

The argument is persuasively, but deceptively, simple. It goes like this. The computer chess program works by exploring all the possible moves as far ahead as possible in a set time. For each move by white, it considers each possible response by black, then for each possible response by black it considers each possible next move by white, and so on. From all these moves, it selects the first move that leads, assuming perfect play, to the most favourable situation for whichever colour it is playing. There are about 10^{120} possible variations in a chess game, and if a computer could analyse a trillion (10^{12}) possibilities every second, it would take 10^{100} years to analyse a whole game. For comparison, the time since the Big Bang is a few times 10^9 years, which is roughly a decimal point followed by 90 zeroes and a 1 times the time needed to analyse the number of possible variations in a chess game. In 1997, using 256 processors working in parallel, Deep Blue examined a mere 60 billion possibilities in the three minutes allowed between moves in the Kasparov game. Each of the processors considered about 25 million possibilities. So a naive understanding of Grover's algorithm suggests that a single quantum processor equivalent to one of Deep Blue's processors could handle 25 million squared possibilities in the same time, making the whole computer 600 million times more powerful than Deep Blue, and able to probe that much further into the web of chess possibilities in the time allowed.

Alas, this naive understanding is wrong. Richard Cleve, of the University of Calgary, pointed out that games like chess, in which there is a relatively small number of fixed choices available at each move, do not lend themselves to speedup by Grover searching, and at best the process would be made slightly faster by a fixed amount. Don't just take my word for it – in a footnote to an article in *Frontiers* magazine in December 1998, David Deutsch said that 'algorithms based on Grover-accelerated brute-force searching' are not suitable for speeding up chess-playing computers significantly – although, of course, there may be other algorithms that are more suited to the game.

As this example demonstrates, there may be limitations on quantum computing that are quite separate from the practical problems of building a quantum computer. To see those limits, it is helpful to look in more general terms at the things computers (not just quantum computers) can do well, and the things they can't do. The key question that has to be considered is how quickly the time needed to solve a problem grows as the size of the problem increases. The time taken for factoring, our classic example, grows exponentially with the number of digits in the number; such problems can be solved efficiently if we have an algorithm that involves a number of steps that grows only by a fixed power, not exponentially, as the number increases, which is why Shor's algorithm works. Such algorithms are said to be 'efficient', and problems which can be solved by efficient algorithms belong to a category mathematicians call 'complexity class P',[20] or are said to be 'in P'. Another class of problem, known as NP,[21] are very difficult to solve but easy to check once you have the answer – such as the factorization problem. All problems in P are, of course, also in NP.

Going a step further, there are so-called NP-complete (or NP-hard) problems. An example is the problem of finding the shortest way to visit on foot all the islands in an archipelago made up of thousands of islands, with every island connected to at least one other island by a bridge, on a tour which visits each island once, without retracing your steps.[22] These are, in a sense, all different versions of the same problem, and it can be proved that if anyone ever found an efficient algorithm solution to one NP-complete problem, it would also solve not just the other NP-complete problems, but any NP problem. This would mean that all NP problems were really P problems, that P equals NP. But if anyone could prove that P is not equal to NP, we would be certain that there is no way to solve an NP problem in a reasonable time on a classical computer. A reward of a million dollars is on offer from the Clay Mathematical Institute in Cambridge, Massachusetts, for anyone who can prove P = NP; but few mathematicians believe that it will ever be won.

Unfortunately, the factoring problem is not NP-complete, so Shor's algorithm does not make him a candidate for the prize; and although it is not possible to prove that there is no quantum algorithm to solve NP-complete problems, there is no evidence that such an algorithm exists. This means that as far as we know, in spite of the hype sometimes associated with them, quantum computers are not going to be able to solve efficiently every interesting problem in the Universe, and it may even be possible to invent quantum codes that are as hard to crack using quantum computers as existing codes are using classical computers. The class of problems that quantum computers can solve efficiently is called BQP (bounded-error quantum polynomial). A quantum computer can solve efficiently any problem that a classical computer can solve efficiently, and certain other kinds of problem also, but it cannot solve efficiently all problems that are intractable to classical computers. Quantum computers are not magic!

And there is another important insight. Because the physics of the Universe is quantum physics, a quantum computer is as good as it gets. Within the known laws of physics, we will never have a better kind of computer, and will never have an efficient way to solve NP-complete problems. We will doubtless develop more powerful quantum computers, just as more powerful classical computers have been developed since the 1940s, but they will only be able to tackle the same kinds of problems that the first generation of quantum computers will be able to tackle. Unless, of course, the laws of physics turn out to need modification in the light of future discoveries. But that way madness lies, and building the first generation of quantum computers is a tough enough task to be going on with.

The Ugly: making it work

Putting all this into practice requires the construction of logic gates operating on quantum principles. These are analogous to the logic gates operating in classical computers, described in Chapter 3, but not identical, because quantum logic is not the same as classical logic. As an example, in a quantum computer it is possible to perform the operation NOT, just as in a classical computer, by poking an electron in an atom, which can exist in one of two states, either 0 or 1, with a judiciously chosen pulse of laser light. The right kind of pulse, known as a π-pulse, will flip the electron from one state to the other – from 0 to 1, or from 1 to 0. Poke it with another π-pulse and you will end up back where you started. But in a quantum computer, you might alternatively poke the electron with an equally carefully selected but weaker pulse known as a $\pi/2$-pulse. This has the effect of putting the electron in a superposition of states 0 and 1. Now, if you poke it with another $\pi/2$-pulse it will settle into the *opposite* state from the one it started out in. Two $\pi/2$-pulses produce the same result as one π-pulse, but there is no classical equivalent of the effect of a single $\pi/2$-pulse. This operation is known as 'Square Root of NOT'. The fundamental feature of this is that if we made a conventional measurement of the electron in its super-position we would get either the answer 0 or the answer 1 with equal probability; using Square Root of NOT we get one answer with 100 per cent certainty. And we have a way of 'negating' a value in a register (changing 0 to 1, or 1 to 0) by doing the same thing twice in succession, which is impossible in a classical computer.

Another variation on the NOT gate turns out to be a crucial component in quantum computation. This is the quantum CNOT (Controlled NOT) gate: a two-bit gate in which the control bit, C, stays the same during an operation, but the second bit changes (from 0 to 1, or from 1 to 0) if C is 1 and stays the same if C is 0. In classical computation, with a set of three-bit gates, such as Fredkin gates, it is possible to build any logic circuit. But much to the relief of experi-menters trying to build a quantum computer, it turns out that a set of quantum CNOT gates, each made up of just two qubits, plus a few single-qubit gates, can be combined to make any logic circuit. Their relief stems from the difficulty of isolating even two qubits and getting them to cooperate in the right way; getting three qubits to work together would be much harder. In quantum computing, but not in classical computing, it is possible to carry out all the roles of three-bit gates using combinations of gates with one or two qubits. The quantum CNOT gate also induces entanglement. I won't go into details, but under the right circumstances a pair of qubits which are initially unentangled can be affected by the gate in such a way that both qubits end up in the same state, without it being determined whether this state is 0 or 1.

There's another feature of the CNOT gate in particular, and quantum computers in general, that is sometimes overlooked. Everything is reversible, as I explained in Chapter 3. It ought to be possible, with a fully operational quantum computer, to reconstruct all the workings that led to a particular answer, and even to determine the question that it answers. If only Douglas Adams were alive to see this. In *The Hitchhiker's Guide to the Galaxy*, a supercomputer gives the answer to 'life, the Universe and everything' as 42, but nobody knows what the question was. If the computer, Deep Thought, were operating on quantum principles, it could simply be asked to run the calculation backwards to reveal the forgotten question!

We are a long way from Deep Thought. But just as the longest journey starts with a single step, the road to a working quantum computer began with the making of a single quantum switch that could be flipped from 0 to 1 and back again; and the experimenters were prodded into taking the first step down this road when Artur Ekert gave a talk about quantum computers to a meeting of atomic physicists held in Boulder, Colorado, in 1994. After spelling out the theory behind quantum computation, Ekert challenged his audience to come up with a single working CNOT gate to demonstrate that the theory could be put into practice. The challenge was taken up by Juan Ignacio Cirac and Peter

Zoller, of the University of Innsbruck. The way they did it provides some insight into the practical difficulties of quantum computation.

Experimenters such as the Innsbruck team already knew how to manipulate single atoms that have been provided with an electric charge – ions. Usually, atoms are electrically neutral, because the positive charge of their central nucleus is balanced by the negative charge of the cloud of electrons that surrounds the nucleus. But it is relatively easy to knock away one of the electrons, leaving a small overall positive charge, or in some cases to encourage an extra electron to stick to the atom, giving an overall small negative charge. In either case, the charge on the resulting ion makes it possible to manipulate it with electric and/or magnetic fields, guiding it into a small vacuum chamber, known as an ion trap. In most of the quantum experiments, such as the ones at Sussex University, the ion is held in place by electric fields, and prevented from jiggling about by laser beams pushing at it from all sides to keep it still,[23] like a big ship being held in place by tugs nudging at it from all sides. Because heat is associated with the random motion of atoms and molecules, this process is known as optical cooling.

The ion trap was invented by Hans Dehmelt, a German-born American physicist working at the University of Washington. He was born in Görlitz on 9 September 1922, and graduated from high school (the Gymnasium zum Grauen Kloster, in Berlin) in 1940. In order to avoid being drafted into the infantry, he volunteered for the anti-aircraft artillery, where he never rose above the rank of senior private but had something of a charmed life in troubled times. He was sent with his battery to relieve the army at Stalingrad, where they were extremely lucky to escape the encirclement, and then, in 1943, was selected under an army programme to study physics for a year at the University of Breslau. Back on active service, he was sent to the Western Front and captured by the American army at the Battle of the Bulge at the end of 1944. After a year as a prisoner of war, Dehmelt studied physics at the University of Göttingen, supporting himself by repairing pre-war radios, all that were available to the civilian population. He completed a Master's degree in 1948 and a PhD in 1950. Two years later, after post-doctoral work in Göttingen, Dehmelt emigrated to the United States, taking up a position at Duke University. He moved to the University of Washington, Seattle, in 1955, staying there for the rest of his career and retiring in 2002. He became a naturalized US citizen in 1961.

Dehmelt had long been intrigued by the idea of isolating and investigating single quantum entities, and in the second half of the 1950s he focused on how electrons might be trapped in this way. His inspiration was a device invented in the 1930s by the Dutch physicist Frans Michel Penning, which used a flow of electrons through a discharge tube in a magnetic field to measure pressure. In a generous gesture, Dehmelt called his device, essentially as described above, a 'Penning trap'; but it was different from Penning's, and it should really be called the Dehmelt trap. Throughout the 1960s, Dehmelt and his colleagues refined such traps until in 1973 they made the breakthrough of trapping a single electron; in 1976 they were able to observe a quantum transition (popularly known as a 'quantum leap') of a single ion held in a trap. For all this (and later) work, Dehmelt was awarded a share of the 1989 Nobel Prize in physics, together with another German physicist, Wolfgang Paul,[24] who had devised a different kind of ion trap (logically, known as the Paul trap), which was also used by the Dehmelt team.

Although the ancillary apparatus needed to operate ion traps is quite large – cooling equipment, magnets, lasers and so on – the traps themselves can be tiny. In the first decade of the twenty-first century, Winfried Hensinger, now at Sussex University, and a team from the University of Michigan made the world's first monolithic[25] ion trap on a microchip, with the ion and the electrode separated by just 0.023 mm, about the width of a human hair. Such traps are made in a similar fashion to conventional computer chips, by depositing an insulating layer, itself made of gallium arsenide sandwiched between layers of aluminium–gallium–arsenide, between two electrode layers. The process, photolithography, produces a three-dimensional 'nano sculpture', chemically etched out of gallium arsenide, consisting of cantilever-shaped electrodes which surround the trapped ion. A channel carved down the chip leads a single cadmium ion into the prepared cavity, where it is cooled and probed using laser light shone down the same channel. Hensinger's team at Sussex has since trapped ytterbium ions in this kind of 'chip compatible' and potentially scaleable setup – indeed, while this book was going to press they produced the world's first two-dimensional lattice on a chip. Similar systems using microwaves rather than optical lasers have also been developed. In an alternative approach, about the same time a group in Boulder, Colorado (of whom more shortly), devised a trap which held up to six magnesium ions in a row, kept hovering above gold electrodes by electrostatic repulsion. A laser beam playing over the surface of the trap provided the cooling and probing.

It's easy to see how, in principle, such a trapped ion could operate as a qubit. A single electron in the outermost region of the ion (the outermost 'shell') could be flipped between two energy states, corresponding to 0 or 1, or placed in a superposition, by pulses of laser light. A row of ions jiggling to and fro in a magnetic field could store information in the qubits of the individual ions, and also in the to and fro motion, which is itself quantized. So a single ion could store 1 from the energy mode and 0 from the rocking mode, giving a two-bit register reading 10, which could be flipped to 11, 01, 00 or a suitable superposition – tantalizingly close to the requirements for a CNOT gate.

The first working CNOT gate was made in 1995 by David Wineland (who had been a member of Dehmelt's team that isolated a single electron in 1973), Christopher Monroe, and other members of the US National Institute of Standards and Technology (NIST) at Boulder, Colorado. It is no co-incidence that they achieved this a year after Ekert's talk in Boulder, although Ekert told me that 'at the time I did not realize that it inspired Dave, Ignacio, Peter and others so much that it effectively marked the beginning of experimental quantum computation'. The NIST team used the rocking motion as one of the qubits, as suggested by Cirac and Zoller; but the second qubit depended not on energy but on the spin of the outer electron. They were able to put single ions of beryllium (Be$^+$) in a state where a laser pulse would change the spin of the electron (from up to down, or vice versa) only if the ion was vibrating in the mode corresponding to 1. In these early experiments the CNOT operation worked properly about 90 per cent of the time, and the ion was maintained in the required state for less than a thousandth of a second before it decohered, too short a time for it to be used in any extended computation. But the point is, it did work. It was a real quantum gate. Since 1995, researchers at NIST and elsewhere have been able to combine up to fourteen ions working together to carry out simple quantum computations. But this is far short of the rows of thousands of ions that would be needed to have a decent crack at factorizing a large number using Shor's algorithm. And the ugly problem of decoherence is made uglier by the need to identify and correct errors as the computation proceeds.

Error correction is one of those processes that gives with one hand while taking with the other, so the trick is to make sure that it gives more than it takes. Think of the steps in a computation as messages. The simplest way to check whether a binary message has been sent correctly is by adding a parity bit. If the number of 1s in the message is even, the parity bit is 0; if the number of 1s is odd, the parity bit is 1. So a seven-digit message 1010010 becomes 10100101 and 1110010 becomes 11100100. If we receive a message 11100101 we know there is a mistake and the message has to be re-sent.

Adding a little more sophistication makes it possible to identify the location of single errors. If the message is sent in lines, each of the same length, arranged in columns, we can add parity bits to both the rows and the columns. So a single error will show up in a single row and in a single column, making it easy to identify. If there are more errors, and if there are errors in the parity bits, the situation gets more complicated, but can still be tackled using this kind of technique. For example, a standard method called the Hamming code, used in classical computing, assigns each of the possible 16 4-bit messages[26] a unique 7-bit codeword chosen so that each of them differs from all of the others in at least three places. This makes it easier to identify errors; but the point I want to emphasize is that, as with all these techniques including the parity bit itself, it does so by adding more bits to the message.

In quantum computation – quantum codes, in the computational, not the cryptographic, sense – there are two additional problems. First, we must be careful not to disturb any of the qubits taking part in the computation, because a measurement, or any observation, would cause decoherence. Secondly, it is impossible to make a duplicate of a set of quantum entities without destroying the original. This is called the 'no-cloning' theorem, and is related to the fact that you cannot measure a quantum system without altering it. You *can* make an exact copy, but you destroy the original in the process. This is the basis of so-called quantum teleportation, when an entity such as a photon in a particular state is probed in one location and the information gleaned is used to construct a photon in the identical state somewhere else (perhaps tens of kilometres away) through entanglement.[27] In effect, the original photon has been teleported from A to B instantaneously. But although that opens mind-boggling possibilities in other areas, it is a major inconvenience at this stage in the development of quantum computing, because one way of checking for errors is to have several copies made of a message and check whether they agree with one another.

One way round these problems would be to add extra qubits which do not take part in the computation to the strings of qubits that are being manipulated inside a quantum computer. Such an extra string of qubits is known as an ancilla, from the Latin word for a maidservant. I've no intention of going into the gory details, but the bottom line is that you need twenty-three ancilla qubits to correct for three simultaneous errors, and eighty-seven of them to correct for seven errors. The obvious question is whether the likelihood of errors in the ancillae outweighs the advantage of having them there at all. But the experts assure us that it will all be worthwhile if the error rate for each component in the computation is reduced to less than 1 in 100,000. Two-qubit gates can now be made that operate with a fidelity of just over 99 per cent, which means that the probability of the gate operating incorrectly is less than 1 in 100. Until recently, it was thought that in order for error correction techniques to work, fidelity would have to be improved to 99.999 per cent. Andrew Steane, of the University of Oxford, describes this as 'daunting but possible'. But in 2011, David Wang, Austin Fowler and Lloyd Hollenberg, of the Centre for Quantum Computation and Communication Technology at the University of Melbourne, showed that error correction could be done with 'only' 99 per cent fidelity. And even though error correction may increase the number of operations required to carry out a computation by a factor of thousands, what does this matter if you have the Multiverse to play with?[28]

But I don't want to end this chapter on such a daunting note. The optimistic news is that quantum computers have already been built and have used Shor's algorithm in the factorization problem. To be sure, only in a modest way; but it's a beginning. In 2001, a team led by Isaac Chuang at the IBM Almaden Research Center in San Jose, California, used a different method of correcting – or rather, compensating for – errors to find the factors of the number 15. The essence of their approach was to work with a molecule which contains five fluorine atoms and two carbon atoms, each of which has its own nuclear spin state.[29] This means that each single molecule can in effect be used as a seven-qubit quantum computer, equivalent to a classical computer with 2^7 bits (128 bits). But they didn't work with just a single molecule. You can't clone a quantum entity into multiple copies of itself, but you can prepare a multitude of quantum entities all in the same state. They used about a thimbleful of liquid containing about a billion billion molecules, probed with pulses of radio-frequency electromagnetic waves and monitored using the technique of nuclear magnetic resonance (NMR) familiar today from hospital scanning systems (where it is known as magnetic resonance imaging, or MRI, because the word 'nuclear' frightens some people).

Left to their own devices, the spins of the nuclei of the atoms in all those molecules 'point' in different directions. For computational purposes they can be regarded as a random string of 0s and 1s. Applying the appropriate magnetic field changes the spins of all the nuclei inside some of the molecules, so that in about one in a hundred million of the molecules all seven nuclei are in the same state – say, 1. That's like 10 billion identical computers all in the state 1111111. More subtle applications of the magnetic influence can flip one particular nucleus in each molecule, so that all 10 billion of them now read, say, 1111011. The pattern can be read by NMR because 10 trillion molecules are giving out the same signal against a background of random noise produced by all the other molecules. And if, say, 10 per cent of the 10 trillion are in the 'wrong' state because of errors, that just gets lost in the noise. The experimenters can flip the spin of each type of atom in all the molecules, effectively at will; but the atoms in each molecule interact with their neighbours, so the molecule used was chosen to have certain properties, so that, for example, one of the nuclei will flip only if its neighbour is in the spin state 1, forming a CNOT gate. Which is how the 'computer' was made to factorize the number 15.

Of course, the experimenters were careful not to read the patterns in the molecules during the computation because that would make them decohere. They only looked at the pattern when the computation was complete. In effect, the 'readout' from the computer averaged over all the molecules, with the huge number of right answers effectively swamping the much smaller number of errors introduced by de-coherence and other difficulties. You will not be surprised to learn that the computer found that the factors of 15 are 3 and 5. But still, it worked. It even made the pages of the *Guinness Book of Records*.

Unfortunately, because of the limitations of the monitoring technique, the method will not work for molecules with more than about ten qubits, so at present there is no prospect of building a larger quantum computer using this method. But there are, as we shall see, realistic prospects of building bigger quantum computers using other techniques.

Chapter Six

Turing's Heirs and the Quantum Machines

Tony Leggett

Everything I have described so far demonstrates that quantum computers work – not just in principle, but at a practical level. Individual qubits can be prepared and manipulated, with the aid of individual logic gates, including the vital CNOT gate. But the enormous challenge remains of constructing a quantum computer on a scale large enough to beat classical computers at the range of tasks, such as factorization of large numbers, for which they are suited. Even given the power of superposition, this will (conservatively) involve manipulating at the very least hundreds of qubits using dozens of gates, within the time limits set by decoherence and with inbuilt error correction. It's a sign of the immaturity of the field that many competing approaches are being tried out in an attempt to find one that works on the scale required. And it's a sign of how fast the field is developing that while I was writing this book a technology that had seemed like an also-ran when I started had emerged as one of the favourites by the time I got to this chapter. I've no idea what will seem the best bet by the time you read these words, so I shall simply set out a selection of the various stalls to give you a flavour of what is going on. The techniques include the trapped ion and nuclear magnetic resonance approaches that we have already met, superconductors and a quantum phenomenon known as the Josephson junction, so-called 'quantum dots', using photons of visible light as qubits, and a technique called cavity quantum electrodynamics (involving atoms).

The key criteria

If any of these techniques is to work as a 'proper' quantum computer, as opposed to the 'toy' versions so far constructed, it will have to satisfy five criteria, spelled out by David DiVincenzo, of IBM's Physics of Information group, at the beginning of the present century:[1]

1   Each qubit has to be well defined ('well characterized' in quantum jargon) and the whole system must be scaleable to sufficiently large numbers. In computer jargon, each qubit must be separately addressable. It is also desirable, if possible, to have a single quantum system acting as different kinds of qubit, as with the single ion we met in the previous chapter that stores 1 from the energy mode and 0 from the rocking mode, giving a two-bit register.
2   There has to be a way of initializing the computer by setting all the qubits to zero at the start of a computation (re-setting the register). This may sound trivial, but it is a big problem for some techniques, including the NMR system that has proved so effective on a small scale. In

addition, quantum error correction needs a continuous supply of fresh qubits in the 0 state, requiring what DiVincenzo calls a 'qubit conveyor belt' carrying qubits in need of initialization away from the region where computation is being carried out to be initialized, then bringing them back when they have been set to 0.

3 There has to be a way of solving the old problem of decoherence, or specifically, decoherence time. A classical computer is good for as long as the hardware lasts, and my wife is not alone in having a computer nearly ten years old that she is still entirely happy with. By contrast, a quantum computer – the virtual Turing machine inside the hardware – 'lasts' for about a millionth of a second. In fact, this is not quite the whole story. What really matters are the relative values of the decoherence time and the time it takes for a gate to operate. The gate operation time may be pushed to a millionth of a millionth of a second, allowing for a million operations before complete decoherence occurs. Putting it another way, during the course of the operation of a gate, only one in a million qubits will 'dephase'. This just about makes quantum computing feasible; or, in DiVincenzo's words, it is, 'to tell the truth, a rather stringent condition'.

4 As I have already discussed, we need reversible gates. In particular, we need to incorporate CNOT gates into the 'circuitry' of the quantum computer, but these gates have short decoherence times and are difficult to construct. (Incidentally, *if* 3-bit gates, equivalent to Fredkin gates, could be made out of qubits, quantum computers would be more efficient; 2-bit gates are the minimum requirement, not the best in computational terms.) It is also

necessary, of course, to turn the gates on and off as required. Quantum computer scientists have identified two potential problems that might be encountered. The first involves gates which are switched on by natural processes, as the computation proceeds, but which are hard to switch off; the second involves gates which have to be switched on (and off) from outside as required. 'Outside' to the gate would be a 'bus qubit' which would have to be able to interact with each of the qubits in the computer, and which would itself be prone to decoherence. But the bottom line is that quantum gates cannot be implemented perfectly, and errors are inevitable. The trick will be to minimize the errors and find ways of working around them.

5   Finally, it has to be possible to measure the qubits in order to read out the 'answer' to a problem. Inevitably, because this involves quantum processes, the measurement cannot give a unique answer with 100 per cent accuracy, so it also has to be possible to repeat the computation as many times as is required to achieve the desired level of accuracy. This may not be too arduous. DiVincenzo points out that if the 'quantum efficiency' is 90 per cent, meaning that the 'answer' is right nine times out of ten, then 97 per cent reliability can be achieved just by running the calculation three times.

DiVincenzo also added two other criteria, not strictly relevant to computation itself, but important in any practical quantum computer. They are a result of the need for communication, by which, says DiVincenzo, 'we mean quantum communication: the transmission of intact qubits from place to place'. This is achieved with so-called 'flying qubits' that carry information from one part of a quantum computer to another part. The first criterion is the ability to convert stationary qubits into flying qubits, and vice versa; the second is to ensure that the flying qubits fly to the right places, faithfully linking specified locations inside the computer.

Nobody has yet found a system which achieves a satisfactory level for all five criteria at the same time, although ion trap devices come closest, having fulfilled all the criteria separately and several in conjunction with one another.[2] Some systems are (potentially) good in one or two departments, others are good in other departments. It may be that the best path will involve combining different techniques in some sort of hybrid quantum computer, to get round all these difficulties. But here are some of the contenders, as of the end of 2012, starting with one of my favourites.

Josephson and the junction

When I first started writing about quantum physics, I was particularly intrigued by work being carried out at Sussex University on Superconducting Quantum Interference Devices, or SQUIDs. These are, by the standards of quantum physics, very large (macroscopic) objects, a bit smaller than the size of a wedding ring. Yet they can behave, under the right circumstances, like single quantum entities, which is what makes them so fascinating. Now, three decades after I first wrote about them, they have the potential to contribute to the construction of quantum computers. They are based on a phenomenon known as the Josephson effect, discovered in 1962 by a 22-year-old student, who later received a Nobel Prize for the work.

Brian Josephson was born in 1940 in Cardiff. He was educated at the Cardiff High School for Boys, at the time a grammar school (it has since merged with two other schools to form the modern comprehensive Cardiff High School). From there he went on to Trinity College, Cambridge at the age of seventeen, graduating in 1960. As an undergraduate he had already published a significant scientific paper concerning a phenomenon known as the Mösbauer effect, and was marked out as a high flier. Josephson stayed on in Cambridge to work for a PhD, awarded in 1964, two years after his Nobel-Prize-winning breakthrough; also in 1962, while still a student, he was elected a Fellow of Trinity. After completing his PhD, Josephson spent a year as a visitor at the University of Illinois before returning to Cambridge, where he stayed for the rest of his career (apart from brief visits to universities around the world), becoming a professor in 1974 and retiring in 2007. But after he encountered Bell's theorem in the mid-1960s, Josephson drifted away from mainstream physics and became increasingly intrigued by 'mind–matter interactions', directing the Mind–Matter Unification Project at the Cavendish Laboratory (a Nobel Prize allows researchers considerable leeway in later life), studying Eastern mysticism, and becoming convinced that entanglement provides an explanation for telepathy. Most physicists regard this as complete rubbish, and feel that Josephson's brilliant mind was essentially lost to physics by the time of the award of his Nobel Prize in 1973.

When the Royal Mail produced a set of stamps to mark the centenary, in 2001, of the Nobel Prizes, they asked laureates, including Josephson, to contribute their thoughts on their own field of study. In his comments, Josephson referred to the possibility of 'an explanation of processes still not understood within conventional science, such as telepathy'. This provoked a fierce response from several physicists, including David Deutsch, who said: 'It is utter rubbish. Telepathy simply does not exist . . . complete nonsense.'[3]

But none of this detracts from the importance of the discovery Josephson made in 1962, which is straightforward to describe but runs completely counter to common sense. He was studying the phenomenon of superconductivity, which had fascinated him since he was an undergraduate. This happens in some materials when cooled to very low temperatures, below their appropriate 'critical temperature', at which point they have no electrical resistance at all. It was discovered in 1911, by Kamerlingh Onnes, in Leiden; but although clearly a quantum effect, it was still not fully understood in 1962.

Josephson found that, according to the equations of quantum physics, under the right circumstances, once started a current would flow for ever through a superconductor without any further voltage being applied. The 'right conditions' involve what have become known as Josephson junctions: two superconductors joined by a 'weak link' of another kind of material, through which electrons[4] can tunnel. There are three possible forms of this junction: first, superconductor–insulator–superconductor, or S-I-S; secondly, superconductor–nonsuperconductor–superconductor, or S-N-S; and finally one with a literally weak link in the form of a thin neck of the superconductor itself, known as S-s-S.

The story of how Josephson came up with his insight has been well documented, notably by Josephson himself in his Nobel lecture, and by Philip Anderson (himself a later Nobel

Prize winner), who was visiting Cambridge from Bell Labs in 1962. Anderson wrote about the discovery of the Josephson effect in an article in the November 1970 issue of *Physics Today*, recounting how he met Josephson when the student – nearly seventeen years his junior – attended a course he gave on solid-state and many-body theory: 'This was a disconcerting experience for a lecturer, I can assure you, because everything had to be right or he would come up and explain it to me after class.' Josephson had learned about experiments involving tunnelling in superconductors, and was working on the underlying theory when 'one day Anderson showed me a preprint he had just received from Chicago in which Cohen, Falicov and Philips calculated the current flowing in a super-conductor–barrier–normal-metal system ... I immediately set to work to extend the calculation to a system in which both sides of the barrier were superconducting.' Josephson discussed his work with Anderson, who was encouraging but, he emphasizes, made no direct contribution: 'I want to emphasize that the whole achievement, from the conception to the explicit calculation in the publication, was entirely Josephson's ... this young man of twenty-two conceived the entire thing and carried it through to a successful conclusion.'

Anderson returned to Bell Labs, where he and John Rowell made the first working Josephson junction and confirmed the reality of the Josephson effect. Meanwhile, in August 1962 Josephson wrote up his work as a 'fellowship thesis' in support of his (successful) application for a research fellowship at Trinity College; this may be a unique example of such a thesis being worthy of a Nobel Prize! There were originally just two copies of this masterpiece, one submitted to Trinity and one kept by Josephson; a third (a photocopy) turned up in Chicago, but Anderson does not know how it got there. More formal publication came in the journal *Physics Letters* later in the year (volume 1, page 251). But in accordance with the regulations of Cambridge University, Josephson still had to remain 'in residence' for another two years before he could be awarded his doctorate.

Josephson's paper was as comprehensive as it could possibly have been. It was clear from the outset that the Josephson effect had many practical applications, some of which I will describe. At Bell, Anderson and Rowell consulted their resident patent lawyer about the possibilities: 'In his opinion, Josephson's paper was so complete that no one else was ever going to be very successful in patenting any substantial aspect of the Josephson effect.' Nevertheless, they took out patents, but these were never enforced so have never been challenged.

Many of the practical applications of Josephson effect devices depend on their extreme sensitivity to magnetic fields – so extreme that in some cases the Earth's magnetic field has to be compensated for. In one of the early experiments by Anderson and Rowling, they found a 'supercurrent' of 0.30 milliAmps in the Earth's magnetic field, which increased to 0.65 mA when the field was compensated for. (*Less* field means *more* current.) Probably the most widespread application is in calibrating and controlling voltage. A Josephson junction exhibits a precise relationship between frequency and voltage. Since frequency is defined in terms of standards such as the emission from atoms of caesium, in effect this leads to a definition of the volt; turning this around, the effect is used to ensure the accuracy of DC voltage standards. Another application, the superconducting single-electron transistor, is a charge amplifying device with widespread potential uses; Josephson devices can also be used as fast, ultra-sensitive switches which can be turned on and off using light, increasing the processing speed of classical computers, for example, a hundredfold. A group at the University of Sussex, headed by Terry Clark, has developed a device based on this technology which is so sensitive that it can monitor a person's heartbeat from a metre away, without the need for any electrical connections; the technique has been extended to monitoring brain activity. This list is by no means exhaustive; but what we are interested in here is the application of the Josephson effect to quantum computing, where the key (as with some of the other applications) is the superconducting quantum interference device, or SQUID. A SQUID is a ring of superconducting material incorporating a single Josephson junction, so that electric current can flow endlessly round the ring without any voltage being applied. But, in an echo of Schrödinger's famous cat 'experiment', the *same* electric current can flow both ways round the ring at once. Quantum superposition can be demonstrated in a macroscopic object, one big enough to see and feel – I've held one in my own hand, courtesy of Terry Clark.

Leggett and the SQUID

My own introduction to SQUIDs came in the 1970s, through my contacts at the University of Sussex, where Tony Leggett, a leading low-temperature physicist and later Nobel Prize winner, was then working. Leggett's unusual route to a Nobel Prize in physics is worth reporting. He was born into a Catholic family in Camberwell, South London, in 1938, the eldest of five children (he had two sisters and two brothers).

The family soon moved to Upper Norwood, where he attended a Catholic elementary school before moving on to the College of the Sacred Heart in Wimbledon. Then, at the end of the 1940s, when his father got a job teaching physics and chemistry at a Jesuit school (Beaumont College) in Windsor, all three of the Leggett boys were allowed free tuition there, and the family moved to Staines, near the modern site of Heathrow Airport. In line with normal practice at the time, at the age of thirteen Tony had to choose which academic path to follow – classics, modern languages, or mathematics and science. As was also the practice at the time, the brightest pupils were steered towards classics, science being regarded as somewhat beyond the pale. So, in spite of his father's position, Tony became a classics scholar, studying Latin and Greek languages and literature. Even here, among the elite, he stood out academically, and was placed in classes with boys two years older than himself, eventually winning a scholarship to Balliol College, Oxford, at the end of 1954. The scholarship was, of course, to study classics (the course usually known as Greats); but in the interval between being awarded the scholarship and going up to Oxford in the autumn of 1955, Leggett was introduced to mathematics by a retired university teacher, a priest who was living at Beaumont – more or less as a hobby to pass the time. Although Leggett was fascinated by the subject and found he had an aptitude for it, maths slid into the background during a happy undergraduate career reading Greats. But near the end of the third year of the four-year course 'it gradually began to dawn on me', says Leggett, 'that I could not go on being a student for ever and must start looking for gainful employment'.

The most desirable possible career seemed to be the nearest thing to a continuation of the student life – a PhD and a university lectureship. The natural choice for someone with Leggett's background would have been philosophy; but he was put off by the fact that, as it seemed to him, there was no objective truth in philosophy, no criterion of whether a piece of work is 'right' or 'wrong'. He wanted, he said, to work in a field where there was 'the possibility of being wrong without being stupid'. The subject which fits that criterion par excellence is physics. With no formal training in the subject, but with the confidence instilled by his experience of advanced mathematics, in the summer of 1958 Leggett applied to do a second Oxford degree, this time in physics, following the completion of his Greats course in 1959. Apart from convincing the powers that be at Oxford that he could cope, there was another not inconsiderable hurdle to surmount. The next year, 1959, was to be the last of compulsory military service in the UK, and as a student Leggett would be exempt; so he had to persuade the draft board that his second undergraduate degree was not merely a ruse to escape the call-up. He is convinced that a major factor in gaining the necessary exemption was the fact that in 1957 the Soviet Union had launched the first artificial Earth satellite, Sputnik 1, and the authorities had at last woken up to the importance of directing the best brains into science and engineering rather than classics (one wonders how many potential good scientists of Leggett's generation were lost by the streaming of the brightest students into the classics). With the draft board convinced, and a new scholarship from Merton College, Leggett commenced his physics course in 1959, emerging with a first class degree and going on to postgraduate work which led to a doctorate in 1964, awarded for investigations into the behaviour of superfluid liquid helium.

By then, he was supported by a fellowship at Magdalen College, which enabled him to spend a year at the University of Illinois at Urbana-Champaign, and a year at Kyoto University in Japan. In Kyoto, he immersed himself in the culture, living in Japanese accommodation, learning the language and avoiding 'foreigners'; this was so unusual in the mid-1960s that, he later learned, his colleagues decided that he must be a trainee CIA agent. After Leggett took up a lectureship at Sussex University (where our paths crossed) in 1967, he continued to travel widely to work at research establishments around the world during the vacations, including another extended visit to Japan after he married a Japanese girl (whom he had actually met at Sussex) in 1972. In 1982, two years after being elected a Fellow of the Royal Society, he was offered and accepted a professorship at Urbana-Champaign, where he was based for the rest of his career. He was awarded the Nobel Prize in 2003 for his contributions to the theory of superconductors and superfluids, and knighted in 2004. 'Above all,' he says, 'I have worked on the theory of experiments to test whether the formalism of quantum mechanics will continue to describe the physical world as we push it up from the atomic level towards that of everyday life,' and he encapsulates this enterprise in his description of a SQUID-based device that acts like 'Schrödinger's cat in the laboratory'.

Common sense would tell us that in a SQUID ring, an electric current could flow one way or the other round the ring, but not both ways at once. Quantum physics says that the ring could exist in a superposition of states, like Schrödinger's cat, one corresponding to a clockwise current and one corresponding to an anti-clockwise flow. This is *not* the same as saying that there are two separate currents, with one stream of electrons going one way and one stream going the other way; the whole ring, a visible, macroscopic object, is in a superposition. Theorists such as Leggett calculated that this situation should produce a measurable effect, described as a 'splitting' of energy levels in the system. At the beginning of the twenty-first century, just such an effect was observed in delicate experiments at the State University of New York at Stony Brook, and at the Technical University of Delft. Subsequent experiments have confirmed the reality of these macroscopic superpositions. As Leggett puts it, this is 'strong evidence for the existence of quantum superposition of macroscopically distinct states'.

All this has profound implications for our understanding of the nature of quantum reality. It suggests that the measurement problem cannot simply be explained (or wished) away by saying that 'collapse of the wave function' occurs just because objects are macroscopic. This might lead to a whole new understanding of quantum reality. But such deep waters are not my concern here. What matters FAPP is that quantum superposition and entanglement involving SQUID rings make them candidates for use in quantum computers.

Computing with SQUIDs

The advantages of using SQUIDs are that both the current and the phase[5] in the ring are quantum entities which can be in a quantum superposition, making them suitable for use as qubits, while different SQUIDs can be entangled with one another. So far, experiments have been done entangling both two and three superconducting qubits, involving simple processors that are in effect solid-state quantum processors resembling conventional computer chips. This makes it possible in principle to build CNOT and other gates. Even better, coupling three qubits is particularly important for some quantum error-correction processes which were demonstrated in a three-qubit system of this kind by Yale researchers in 2012. SQUIDs have the huge advantage over atomic-scale systems that they can be 'engineered' in a more or less standard, classical way, and manufactured in large numbers on chips using existing technology (they don't *have* to be as big as wedding rings, of course). A team at the University of California, Santa Barbara, has already managed to put nine Josephson-junction-based quantum devices on a single 6 mm by 6 mm chip, although this does not in itself function as a quantum computer.

The disadvantages are that neither the superpositions nor the entanglements last long (nothing unique about that), and the whole thing has to be operated at very low temperatures, close to absolute zero (−273 degrees Celsius) at 0.8 degrees on the Kelvin scale. But researchers have cracked one of the key DiVincenzo requirements of a quantum computer, by using SQUIDs to develop the quantum bus that is needed to interact with each of the qubits in the computer.

The technique involves placing two SQUIDs in a cavity between two layers of conducting material, where microwave photons (just like the photons in a microwave oven) can bounce between the conductors. The SQUIDs can emit and absorb photons, and the way they do this is tuned by adjusting the voltage across the gap. A SQUID that absorbs a photon becomes 'excited', and if it is already excited it can be triggered to emit a photon into the cavity. This process is a specific example of the phenomenon known as resonance, and is sometimes referred to as 'hybridization' of the qubit and photon states. In resonance, the SQUID switches back and forth between excited and non-excited states. In order to transfer a quantum state from one SQUID to the other, the SQUIDs are first both tuned 'off resonance' and one is put in some specified quantum state. This SQUID is then tuned in resonance to the cavity, and at the appropriate moment when the quantum rules tell us that the probability of its being excited is zero, it is tuned off resonance. This leaves the photon that 'belongs' to the SQUID bouncing around in the cavity. The second SQUID is then tuned into resonance, and interacts with the photon left behind by the first SQUID. At the appropriate moment this SQUID is again tuned off resonance. At that point, the quantum state of the first SQUID has been transferred to the second SQUID. This is known as 'quantum optics on a chip'. It's a small step – transferring a state between a single pair of SQUIDs. But it has been done, and it is a step in the right direction, suggesting, perhaps, a specific role for SQUIDs in computers that may also incorporate other kinds of qubit.

Perhaps significantly, IBM is making a major effort to develop superconducting quantum computer technology. Asked how long it will be before we see a practicable quantum computer, Mark Ketchen, manager of the Physics of Information group at IBM's Watson Research Center, said in 2012: 'I used to think it was 50 [years away]. Now I'm thinking it's 15 or a little more. It's within reach. It's within our lifetime.'[6]

But this is not the only game in town. In a parallel development, so-called quantum dots have also been inserted into microwave cavities and probed in the same way, by a team at Princeton. This, again, is an eminently scaleable technology, but starting from a very different scale.

Corralling with quantum dots

At the other end of the physical scale from macroscopic superconducting qubits, we find the possibility of using single electrons as the bits in a quantum computer. This is the ultimate development of conventional computer chip technology using semiconductors, where a well-developed industry already exists based on the ability to 'build' structures on the scale of nanometres (one nm is a billionth of a metre). The construction process involves depositing layers of semi-conducting material, one on top of the other, in so-called semiconductor lithography, and interesting things happen where layers of different material meet.

The basic idea is to create a three-dimensional structure, like a submicroscopic bubble, in which a single electron can be 'corralled' – confined in a small volume with a known energy level – and can be moved up and down energy levels as required. These corrals are known as 'quantum dots'. The sizes (diameters) of quantum dots are in the range from 5 nm to 50 nm; they can form spontaneously when one semi-conductor material is deposited on another layer of a different semiconductor, because the different electrical properties of the two layers can cause atoms to migrate in a thin layer parallel to the boundary between the two materials, forming patterns in which the overall effect of the electric field of the atomic nuclei creates dips in the local electric field. These dips can be thought of as the electrical equivalent of potholes in a worn road. Electrons can be trapped in these (three-dimensional) potholes in a similar way to pebbles being trapped at the bottom of a pothole.

Another approach, pioneered at the University of New South Wales, manipulates individual atoms to form the quantum dot. A team led by Professor Michelle Simmons made a quantum dot by replacing seven atoms in a silicon crystal with phosphorus atoms. The dot is just 4 nm across, and acts as a classical transistor. Even without quantum effects, this offers the prospect of smaller, faster computers; but 'we are basically controlling nature at the atomic scale,' says Simmons, and 'this is one of the key milestones in building a quantum computer'.[7]

But what can you do with such a trapped electron? There are two particularly promising possibilities as far as quantum computing is concerned. The first is known as the charge qubit. It involves two neighbouring quantum dots, and an electron which can be moved from one quantum dot to the other.[8] Or, of course, it can be in a superposition, where no decision about which dot it is in can be made until it is measured. If the dots are labelled left (L) and right (R), then we have a qubit with L corresponding to 0 (say) and R corresponding to 1. Two-qubit systems built along these lines have already been constructed; one of the most promising possibilities, potentially scaleable, involves two layers of gallium arsenide (a material already widely used in semiconductor technology) separated by a layer of aluminium–gallium–arsenide. This ticks one of the DiVincenzo boxes; another plus is that initialization can easily be achieved by injecting electrons into the system. On the negative side, decoherence times are short, and although two-qubit systems have been constructed, it has not yet been possible to make a two-qubit gate out of quantum dots.

So how about using spin? The fairly obvious idea of representing the state of a qubit by the spin of a single electron, reminiscent of the way the NMR technique uses the spin of an atomic nucleus, has been given the fancy name spintronics. Spin has the great advantage for computer work that it is a very clear-cut property.[9] There are two, and only two, spin states, which can be thought of as 'up' and 'down', and which can be in a superposition. Techniques which define the state of a qubit in terms of energy, for example, may specify the two lowest energy levels of a system as '0' and '1', but there are other energy levels as well, and electrons (or whatever is being used to store the information) can leak away into those levels. Decoherence times for electron spins can be as long as a few millionths of a second, but because electrons have such a small mass it is easy to alter their spin state using magnetic fields. This is particularly important in manufacturing fast gates. So spintronics is a promising way of satisfying the DiVincenzo criterion of having gates that operate faster than decoherence occurs. Even better, it ought to be possible to store information using the spins of nuclei, which have still longer decoherence times, up to a thousandth of a second, and then to convert the information into electron spins for processing.

As with the charge qubit, the technique ought to be scaleable – it is probably the easiest of all the techniques I am describing to scale – and simple to initialize; measurement is not too difficult. But, as ever, there is a downside. Until recently, nobody was able to address the spin of a single electron. This would make it impossible to construct a set of quantum gates. And in order to minimize the decoherence problem, such systems, like the superconducting systems, have to be run at very low temperatures, close to absolute zero. But the Princeton experiments involving quantum dots trapped in microwave cavities offer the potential for addressing individual electrons, with in effect the properties of an electrical system a centimetre or two long being determined by the spin of a single electron.

It's a sign of how fast work in this field is progressing that in the very week I was writing this section another team at the University of New South Wales announced that, in collaboration with researchers from the University of Melbourne and from University College, London, they had succeeded in manipulating an individual electron spin qubit bound to a phosphorus donor atom in a sample of natural silicon.[10] They achieved a spin coherence time exceeding 200 microseconds, and hope to do even better in isotopically enriched samples. 'The electron spin of a single phosphorus atom in silicon', they say, 'should be an excellent platform on which to build a scaleable quantum computer.' Their planned next step – probably achieved by the time you read this – is to manipulate the spins of electrons associated with two phosphorus atoms about 15 nm apart; the electrons have overlapping 'orbits', and the spin imparted to the electron on one atom will depend on the spin of the electron associated with the other atom. This is the basis of a two-bit gate.

In such a fast-moving field, it would be foolish to try to keep up with the pace of change in a book that will not be published for several months after I finish writing; so, as ever, I will largely restrict myself to the basic principles of the different techniques. The one that bears most resemblance to the electron spin approach uses nuclear spin; not in the way we encountered it before, but as one of the most promising prospects for quantum computer memory.

The nuclear option

In 2012, news came of two major developments in nuclear-spin quantum memory, reported in the same issue of the journal *Science*.[11] Both are based on the kind of solid-state technology familiar to manufacturers of classical semi-conducting computer chips. The first involves ultra-pure samples of silicon-28, an isotope which has an even number of nucleons (protons and neutrons) in each atomic nucleus, so that overall there is no nuclear spin. The samples are 99.995 per cent pure. This provides a background which the researchers, a team from Canada, Britain and Germany, describe as a 'semiconductor vacuum'. It has no spins to interact with the nuclei of interest, which greatly reduces the likelihood of decoherence. With this material as a back-ground, the silicon can be doped with donor atoms such as phosphorus[12] (just as in conventional chips), each of which does have spin.

The tern 'donor' means that the phosphorus atom has an electron which it can give up. Each silicon atom can be thought of as having four electronic bonds, one to each of its four nearest neighbours in a crystal lattice; an occasional phosphorus atom can fit into the lattice, also forming four bonds with its neighbours, but with one electron left over. Such a doped silicon lattice forms an n-type semiconductor. Using a coupling which is known as the hyperfine interaction, the spin state of the donor electron (which can itself, in principle, act as a qubit) can be transferred to the nucleus of the phosphorus atom, stored there for a while, then transferred back to the electron. All of this involves manipulating the nuclei with magnetic fields, running the experiment at temperatures only a few degrees above absolute zero, and monitoring what is going on using optical spectroscopy. But crucially, although they appear daunting to the layman, the techniques used to monitor hyperfine transitions optically are already well established as standard for monitoring ion qubits in a vacuum.

In the experiments reported so far, ensembles of nuclei, rather than individual phosphorus nuclei, were monitored. But the decoherence time was 192 seconds, or as the team prefer to point out, 'more than three minutes'. We have already arrived in the era of decoherence times measured in minutes, rather than seconds or fractions of a second, which is a huge and valuable step towards a practical working quantum computer. And the technique should be extensible to the readout of the state of single phosphorus atoms, as well as being suitable to other donor atoms.

Compared with this, the achievement of the other team reported in the same issue of *Science* may seem at first sight less impressive. Using a sample of pure carbon (essentially, diamond) rather than silicon, a joint US–German–British team achieved a decoherence time of just over one second. But they did so at room temperature, reading out from a single quantum system, and they make a reliable estimate, based on cautious extrapolation from the existing technology, that 'quantum storage times' exceeding a day should be possible. That really would be a game changer.

In these experiments, crystals of diamond made from

99.99 per cent pure carbon-12 (which, like silicon-28, has no net nuclear spin) were grown by depositing them from a vapour. Like a silicon atom, each carbon atom can bond with four neighbours. But such crystals contain a few defects known as nitrogen-vacancy (N-V) centres. In such a defect, one carbon atom is replaced by a nitrogen atom, which comes from the air; but since each nitrogen atom can only bond with three carbon atoms there is a gap (the vacancy) where the bond to the fourth next-door carbon atom ought to be. In effect, this vacancy contains two electrons from the nitrogen atom and one from a nearby carbon atom, which exist in an electron spin resonance (ESR) state. N-V centres absorb and emit red light, so they interact with the outside world and can be used as readouts of the quantum state of anything they interact with at the quantum level – the fifth of the DiVincenzo criteria – or as a means of making inputs to the system. The bright red light associated with N-V centres also makes it easy to locate them in the crystal.

The 'anything' the N-V centre interacts with in these experiments is a single atom of carbon-13, which has overall nuclear spin, located one or two nanometres away. At this distance, the coupling between the carbon-13 nucleus and the ESR associated with the N-V centre (another example of the hyperfine interaction at work) is strong enough to make it possible to prepare the nucleus in a specified quantum spin state and to read the state back, but not strong enough to cause rapid decoherence. For the concentration of carbon-13 used in the experiments, about 10 per cent of all the naturally occurring N-V centres had a carbon-13 nucleus the right distance away to be useful; but each measurement involved just a single N-V centre interacting with a single carbon-13 nucleus. Other experiments have shown that it is possible to entangle photons with the electronic spin state of N-V centres, providing another way of linking the nuclear memory to the outside world, potentially over long distances.

The storage time achieved in these experiments was 1.4 seconds. But even using simple refinements, such as reducing the concentration of carbon-13 to decrease the interference caused by unwanted interactions, it should be possible to extend this by more than 2,500 times, to an hour or so; from there it will be a relatively smaller step, using techniques pioneered in other fields, to go up by another factor of 25 to get nuclear spin memories that last for more than a day, at room temperature. But proponents of quantum computing are still far from putting all their eggs in one basket, attractive as this one might be. Even the NMR approach, which is now almost ancient history by the standards of the field, is still providing potentially useful insights.

The nuts and bolts of NMR

In the previous chapter, I got a little ahead of myself by describing the exciting first results of quantum computation using nuclear magnetic resonance, the first successful quantum computation technique, without really explaining the fundamental basis of NMR. It is time to redress the balance.

Atomic nuclei are made up of protons and neutrons.[13] The simplest nucleus, that of hydrogen, consists of a single proton; the next element, helium, always has two protons in the nucleus, but may have either one or two neutrons; these varieties are known as helium-3 and helium-4, respectively, from the total number of particles (nucleons) in the nucleus.

Going on up through heavier chemical elements, the very pure form of silicon used by Michelle Simmons and her colleagues is a variety (isotope) known as silicon-28, because it has 28 nucleons (14 protons and 14 neutrons) in the nucleus of each atom. Another isotope, silicon-29, has 14 protons and 15 neutrons in each nucleus. The crucial distinction, for the purposes of quantum computation, is the difference in spin between nuclei with an even number of nucleons and nuclei with an odd number of nucleons.

Neutrons and protons are both so-called 'spin-½' particles. This means that they can exist in either of two spin states, $+½$ or $-½$, also known as 'up' and 'down', which can be equivalent, as we have seen, to 0 and 1 in binary code. You might think that this would mean that the overall spin of a nucleus of silicon-28 would be anything up to 14, depending on how the spins of individual nucleons add up or cancel out; but the quantum world doesn't work like that. Instead, each pair of protons aligns so that the spins cancel out, and the same is true for each pair of neutrons. So nuclei with even numbers of both protons and neutrons have zero overall spin, but other nuclei have non-zero spin. Thus silicon-28 has zero spin, but silicon-29 has an overall spin of $±½$. This is what makes pure silicon-28 such a perfect background material against which to monitor the spins of atoms used to dope the crystal lattice.

NMR, though, doesn't use nuclei as complicated as those of silicon-28. It depends on the fact that there is an interaction between magnetism and nuclear spin, so that applying the right kind of alternating magnetic field to a nucleus can make it jump between energy levels corresponding to different spin states. This is the resonance in nuclear magnetic resonance, and it shows up as an absorption of energy at a precise frequency of oscillation, the resonance frequency. The simplest nucleus to work with is the hydrogen nucleus, which is a single proton. The exact response of the proton to the oscillating magnetic field depends on its chemical environment – which molecules the hydrogen atoms are part of – so by sweeping a varying magnetic field across the human body and measuring the resonance at different locations it is possible to get a map which reveals the chemical environment of the hydrogen atoms in different parts of the body. That's what we know as an MRI scan.

The curious feature of quantum computing using NMR, though, is that we are dealing not with individual spins, but with some kind of average of billions and billions of essentially identical states – typically involving 10^{20} nuclei. In a fluid[14] being used for quantum computation in this way, the energy difference between the two spin states of the proton is very small, and this means that although nuclei prefer to be in the lower energy level, it is easy for them to get knocked up into the upper level by random interactions (literally, by neighbour atoms bumping into them). Once there, they will fall back down again; but meanwhile, other nuclei have been bumped up to take their place. At any one time, for every million nuclei in the upper level there may be only a million and one in the lower energy level. In effect, the NMR computing technique is working with the one in a million 'surplus' nuclei in the lower level, getting them to jump up to the higher level. But it is working with *all* of them at once. And all of those 'one in a million' nuclei, billions of them, jumping together between energy levels, have to be regarded as a single qubit, switching between the states 0 and 1.

I explained in the previous chapter how effective the technique has been in demonstrating the techniques of quantum computing with small numbers of qubits (up to 10 or so). But as I also mentioned, there are severe scaling problems with the technique, and it has already been pushed about as far as it can go. Even so, it isn't quite ready to be consigned to the dustbin of history. There is something very odd about NMR computation, which has set people thinking along completely different lines – as I shall discuss in the Coda.

Meanwhile, another old warhorse of a quantum computation technique, the ion trap approach, which *is* scaleable, has quietly made steady progress. This was recognized in 2012 by the award of a half-share of the Nobel Prize in physics to David Wineland, of NIST, whom we met in Chapter 5.

Trapped ions take a bow

Wineland's Nobel citation specified that the award was for 'ground-breaking experimental methods that enable measuring and manipulation of individual quantum systems'. In the words of the Royal Swedish Academy of Sciences, which administers the awards, if the quantum computer is built in the near future it 'will change our everyday lives in this century in the same radical way as the classical computer did in the last century'. In this connection, there is one important feature of the ion trap approach which should be borne in mind when considering all the possibilities for computing with quanta. It is the only method in which all of the physics involved uses standard techniques that have all been tried and tested and proved to work. Every one of the other approaches, even though they are based on sound theoretical principles, relies on there being some kind of practical physics breakthrough in the not too distant future if they are to maintain momentum. So while one or another of them may seem to spurt ahead for a time, like the fabled hare, so far they have each ground to a halt after a while, while the trapped ion technique continues to plod along, tortoise-like, improving the technology but always using the same physics.[15] Winfried Hensinger says that the first working quantum computer, in about the middle of the 2020s, is likely to be based on the trapped ion technique, and to be as big as a house. But you only have to compare the size of Colossus with the size of a modern smartphone to realize that this will be far from the end of the story.

Wineland has helped the trapped ion tortoise to take a few more steps down that road. Born in Milwaukee, Wisconsin, on 24 February 1944, he moved to California as a child and attended high school in Sacramento. He took his first degree at the University of California, Berkeley, graduating in 1965, received his PhD from Harvard in 1970, and worked at the University of Washington in Hans Dehmelt's group before joining the National Bureau of Standards in 1975. One focus of his work with trapped ions there has been towards the development of more accurate clocks – better timekeeping devices than the atomic clocks which are now standard. In 1979 Wineland founded the ion storage group of the Bureau, now based at NIST in Boulder, Colorado, using the techniques which I described in the previous chapter.

As I have explained, the problem with developing the trapped ion technique into a practical quantum computer is that it is extremely difficult to control strings of trapped ions containing more than about 20 qubits. Wineland and his colleagues have proposed getting round this difficulty by dividing the 'quantum hardware' up into chunks, carrying out calculations using short chains of ions that are shuffled about on a quantum computer chip by electric forces which do not disturb the internal quantum states of the strings. According to Wineland and Monroe,[16] 'the resulting architecture would somewhat resemble the familiar charge-coupled device (CCD) used in digital cameras; just as a CCD can move electric charge across an array of capacitors, a quantum chip could propel strings of individual ions through a grid of linear traps'. In 2005, Hensinger, at the University of Michigan, and his team, managed to demonstrate reliable transport of ions 'round the corner' in a T-shaped ion trap array. Since then, even more complicated ion trap CCDs have been developed.

This is an active line of research today at NIST, where the researchers work with beryllium ions. Although the electrodes used to guide the ions in a practicable quantum computer would have to be very small – perhaps as little as 10 millionths of a metre across – Monroe and Wineland emphasize that the engineering involved uses just the same kind of microfabrication technologies that are already used in the manufacture of conventional computer chips. Other groups are also working along these lines, undaunted by the need to reduce noise by cooling the electrodes with liquid nitrogen or even liquid helium. But there is another way to combine information from different strings of ions in a quantum computer – using light.

In this approach, instead of using the oscillatory motion of the ions (or ion strings), photons are used to link the qubits together. Ions emit light, and it is possible to set up situations in which the properties of the emitted photons, such as their polarization or their colour, are entangled with the internal quantum states of the ion that is doing the emitting. Photons from two different ions are directed down optical fibres towards a device like the beam-splitting mirrors I described in Chapter 5, but working in reverse. With this setup, the photons enter the 'splitter' from opposite sides, and are given the opportunity to interact with one another. If they have the same appropriate quantum property (the same polarization, for example), they will interact with one another, become entangled, and leave the beam splitter together along the same fibre optic. But if they have different quantum properties – different polarizations, or different colours, or whatever – they will ignore one another and leave the splitter along different optical fibres. Simple photon detectors placed at the end of each fibre tell the experimenters whether entanglement has occurred or not. Crucially, though, there is no way to determine which ion has emitted which photon; but if the detectors reveal that the photons are now entangled with one another, the ions they came from have also become entangled. Although ion–photon entanglement is tricky to work with, the incentive is that it allows for the possibility of a modular quantum ion processor, built up from many smaller processors linked by photons. Eventually, this could lead to a quantum internet.

As is so often the case with quantum experiments, most of the time the emitted photons are never gathered up by the beam splitter, and the entanglement does not occur. But, as ever, the solution is simply to keep trying until the experimenters do find photons being detected simultaneously at the appropriate detectors. Once the detectors show that there is entanglement between the two ions – the two qubits – the experimenters also know that manipulating one of the qubits will affect the other one – the basis of the CNOT gate. This is not just abstract theorizing. A team at the University of Michigan who later moved to the University of Maryland have successfully entangled two qubits in this way in the form of trapped ions separated by a distance of roughly a metre. This is Einstein's 'spooky action at a distance' put to practical use.

In these first experiments, the rate at which ion pairs were entangled was only a few per minute. There is a possible way to make the process more efficient by surrounding each ion by highly reflective mirrors, to make what is known as an optical cavity in which photons bounce around before being trapped in an optical fibre. The technology is tricky; but intriguingly it is closely related to the work for which the other half of the 2012 Nobel Prize in physics was awarded, to the Frenchman Serge Haroche, a good friend of Wineland who was born in the same year as him, 1944.

But before I describe the work for which Haroche received the Nobel Prize, a little diversion, into the world of quantum teleportation. It sounds like science fiction, but it is sober science fact; and it turns out to be highly relevant to one of the most promising approaches to making computing with quanta practicable.

The teleportation tango

Quantum teleportation is based on the spooky action at a distance that so disgusted Einstein but is demonstrated to be real in tests of the EPR 'paradox' and measurements of Bell's inequality. It rests on the fact – confirmed in those experiments – that if two quantum entities, let's say two photons, are entangled, then no matter how far apart they are, what happens to one of those two photons instantly affects the state of the other photon. The key refinement is that, by tweaking the first photon in the appropriate way (called a 'Bell-state measurement'), its quantum state can be transferred to the second photon, while the state of the first photon is, of course, changed by being tweaked. In effect, the first photon has been destroyed and the second photon has become what is termed in common parlance a clone of the first photon. Since the original has been destroyed, however, for all practical purposes the first photon has been teleported to the location of the second photon, instantly. It is *not* a duplication process (and it has also been done with trapped ions!).

There's one small catch. In order to complete the transformation, information about the way the first photon was tweaked has to be transmitted to the location of the second photon by conventional means, no faster than the speed of light. This information is then used to tweak the second photon in just the right way (*not* the same way that the first photon was tweaked, but in a kind of reverse process) to complete the transformation. In effect, the conventional signal tells the system what tweak has been applied to photon number one, and the system then does the opposite to photon number two. Quantum teleportation requires both a quantum 'channel' and a classical 'channel'; it takes two signals to dance the teleportation tango.

A large and successful research effort has gone into making this reality, not least because quantum information offers a way of transmitting information utterly securely using systems that cannot be cracked. I have explained the details in my book *Schrödinger's Kittens*, but the essential point is that information travelling by the quantum 'channel' cannot be read by a third party; in addition, any attempt to eavesdrop will alter the quantum state of the photons, making it obvious that they have been interfered with. This is not the reason why teleportation helps in the design of quantum computers; indeed, in recent times headline-making developments in quantum teleportation have concentrated on much larger scales than those appropriate for computation. But their success emphasizes the reality of the process, and how good scientists now are at working with quanta.

In 2012, two record-breaking experiments made those headlines – both of which will probably have been superseded by the time you read this. First, a large group of Chinese researchers succeeded in teleporting a quantum state through 97 kilometres of open air across Qinghai Lake, using a telescope to focus the photons. Almost as an aside, the experiments confirmed the by-now-expected violation of Bell's inequality, offering insight for the theorists into the foundations of quantum physics. A few weeks later, a team from Austria, Canada, Germany and Norway teleported the properties of a photon across a distance of 143 kilometres, from the astronomical observatory at La Palma, in the Canary Islands, to a European Space Agency ground station on the neighbouring island of Tenerife. Both the transmitting station and the receiving station were located roughly 2,400 metres above sea level, where the air is thin and atmospheric interference is reduced.

But the air is even thinner at higher altitudes, so that in some ways it should be easier to carry out quantum teleportation, and achieve secure communication, by pointing the beams upward to a satellite. The distances involved are very similar to those already achieved on the ground, and although there are, of course, many other problems involved in establishing this kind of satellite communication, the Chinese are already planning a satellite experiment, provisionally scheduled for launch in 2016 or 2017, to test the possibilities, using ground stations in Europe and in China to communicate with the satellite simultaneously for a few minutes in each orbit.[17] This is particularly important because this kind of quantum information is soon lost if the photons are sent through fibre optic cables. The leader of the Chinese team, Pan Jianwei, of the University of Science and Technology of China in Hefei, envisages an eventual network of satellites acting as repeater stations for global coverage of a quantum communications network. This could be the basis of an utterly secure quantum internet; and in all probability many of the computers plugged into that internet will by then themselves be running on quantum principles, including teleportation.

In connection with this work, Chinese researchers have devised ever better techniques for entangling photons. In 2004, they could produce a few four-photon entanglement events every second; by 2012, they could produce entangled groups of four photons at a rate of a few thousand per second. This is important for the communications work, but also, as I shall shortly explain, for some kinds of quantum computing. It's time to return to the main thread of my story.

Fun with photons[18]

Serge Haroche was born on 11 September 1944 – making him just seven months younger than David Wineland – in Casablanca, Morocco, which was then a French 'protectorate'. It did not become fully independent until 1956, at which point the Haroche family (his father was a lawyer and his mother a Russian-born teacher) left for France. He graduated from the École Normale Supérieure in Paris in 1967, and received his PhD from the Pierre and Marie Curie University of Paris in 1971. While working for his doctorate, Haroche was a research associate at the National Centre for Scientific Research (CNRS, from the French title), where he stayed as a research fellow from 1971 to 1973 (including a year, 1972/3, as a visitor at Stanford University) and as a senior research fellow from 1973 to 1975. In 1975 he was appointed a professor at the University of Paris, moving in 2001 to the Collège de France, where he remains, as Professor of Quantum Physics. As well as his first Stanford trip, Haroche has at various times been a visiting scientist at Stanford, Harvard, Yale and MIT.

The common thread of Wineland's and Haroche's work is that both of them make direct observations of individual quantum systems without disturbing their quantum states. But they approach the task from opposite directions. Wineland traps individual ions and both manipulates and monitors their behaviour with light; Haroche, a pioneer of the technique known as cavity quantum electrodynamics (CQED), uses atoms to manipulate and monitor the behaviour of trapped photons. He actually uses microwave photons, with wavelengths longer than those of visible light, but to a physicist all photons are 'light'.

The way to trap a photon is with mirrors. In Haroche's lab in Paris, hollow half-spherical mirrors made of super-conducting material are placed about 3 centimetres apart, making a hollow cavity, and cooled to less than one degree above absolute zero. These mirrors are so reflective that a photon can bounce back and forth between them for 130 milliseconds (more than a tenth of a second) before it is absorbed. Since photons travel at the speed of light, just under 300,000 kilometres per second, this means that an individual photon travels roughly 40,000 kilometres, back and forth across the same 3 centimetres, before it is absorbed. This is approximately equivalent to flying once around the equator of the Earth. While it is doing so, Haroche manipulates it using rubidium atoms in a specially prepared state, known as Rydberg atoms after the Swedish physicist Johannes Rydberg, a nineteenth-century pioneer of atomic spectroscopy. In a Rydberg atom, the outer electrons have been given an energy boost which has lifted them up into 'orbits' much farther out from the nucleus and the inner electrons than usual. They may be as much as 125 nanometres across, a thousand times bigger than ordinary atoms, and interact strongly with microwave photons through a phenomenon known as the dynamical Stark effect. A Rydberg atom sent through the cavity at a carefully controlled speed just has time to interact with the photon, inducing a phase shift in its quantum state – in effect, a change of sign but not a change of size. The photon is unaffected by this, but in the process has become entangled with the Rydberg atom. By analysing the state of the Rydberg atom after it emerges from the cavity, and comparing it with the state it was in when it entered the cavity, Haroche and his colleagues can determine the state of the photon, without it decohering. In an extension of this technique, they can count the number of photons in the cavity, which is much harder than you might guess.

Using these techniques, Haroche and his colleagues have been able to put the photons into a superposition (a 'Schrödinger's cat state') in which, in effect, they are equivalent to waves going in opposite directions at the same time, then to monitor them with Rydberg atoms to see how long it takes for the superposition to decohere. With feedback processes being used to preserve the 'cat state' for longer, this is one possible route to a qubit based on light. But other teams are also having fun with photons, and may be closer to a breakthrough in the field of what is now called quantum photonics.

With their love for acronyms, physicists sometimes refer to the processes involved in quantum computation as QIP, for Quantum Information Processing. Single photons can be manipulated relatively easily, using properties such as polarization, to act as single qubits; but, as we have seen, a key step in QIP is the ability to manipulate pairs of qubits to act as CNOT gates. This involves 'flipping' the state of a target qubit (T) only if the control qubit (C) is in the state 1. The way this is achieved in what has become known as linear optical quantum computing is for the control and target qubits (photons) to be shepherded through an optical network of mirrors, half-silvered mirrors, and so on, together with two other photons. The conditions for the CNOT operation to occur exist within the network (only for the C and T photons), but the experimenters only know if it has taken place when the photons emerge. There are two detectors at the output end of the business part of the network, and in this case if a single photon is detected at each output then the CNOT operation has taken place. This happens only about one-sixteenth of the time – as ever in the quantum world we have to deal with probabilities, not certainties. This is unfortunate, because the probability of a successful computation decreases exponentially with the increase in number of

CNOT gates, making scaling impractical. But there is a way out.

Now comes the clever bit – and it involves quantum teleportation. Remember those two extra photons, passing through the network without being involved in the CNOT operation? The mechanics of the process are complicated, but it is possible to entangle each of the 'working' photons with one of the 'spare' photons, and to combine this with a teleportation process. Within the network, the CNOT operation is attempted repeatedly, until the operation is successful, and only then does the result appear as output. In terms of classical physics, this seems like pure quantum magic, and almost like time travel – the control and target qubits are teleported onto the output photons only after it is known that the gate has succeeded. The teleportation technique could also be applied in reading data from other kinds of quantum computer, such as those based on trapped ions.

I have described the basics of linear optical quantum computing in terms of conventional optical components such as mirrors and beam splitters, and this is indeed the way the first CNOT gate for single photons was built, by Jeremy O'Brien and his colleagues at the University of Queensland, in Australia. It took up an entire laboratory bench, a couple of square metres in area, with photons propagating through the air. Fine for demonstrating that the optical CNOT operation could be made to work, but hardly practicable for scaling up to a working quantum computer containing hundreds or thousands of gates. But Colossus was also based on glass technology – valves – and thanks to miniaturization using semiconductors, we now have computers far more powerful than Colossus, containing far more gates, that we can hold in the palm of a hand. Classical computing had to wait for semi-conductor technology to be developed before the computers could be miniaturized; but Turing's heirs, such as O'Brien, already have the semiconductor technology, enabling them to move straight from the lab bench 'proof of principle' to miniaturization. That's just what O'Brien has done, now working at the University of Bristol, in England, where he is director of the Centre for Quantum Photonics.

By 2008, O'Brien and his colleagues had developed a device containing hundreds of CNOT gates in a piece of silicon just 1 millimetre thick. Instead of mirrors and beam splitters, the device steers photons through a network of waveguides each a millionth of a metre across: channels of transparent silica grown onto a silicon base using standard industrial techniques.[19] Indeed, these 'chips', just 70 mm by 3 mm, were manufactured by industry – CIP Technologies, of Ipswich, in Suffolk – not by university technicians. The 'silica on silicon' technology is widely used in optical telecommuni-cation devices, where the silica guides light in the same way as optical fibres but on a smaller scale. In the pioneering Bristol device, four photons are guided into the network and are put into a superposition of all possible four-bit inputs; the calcu-lation performed by gates inside the network creates an entangled output, which is collapsed by measuring the output states of the appropriate pair of photons. In this way, the Bristol team used Shor's algorithm to determine the factors of 15, proudly finding the answer 3 and 5. All done at room temperature in a device superficially similar to a common computer chip.

'These results show', said the team in an article in *Science*,[20] 'that it is possible to directly "write" sophisticated photonic quantum circuits onto a silicon chip, which will be of benefit to future quantum technologies based on photons, including information processing.'

Although 'all' you need to do to tackle bigger problems is to add more qubits, the 'adding more qubits' step is at present a major hurdle, since it means finding a reliable source of single photons. But there is no reason to think this is beyond the technological capabilities of an industry that has already developed the conventional microchip. Another possible approach is to combine this technique with one of the other techniques I have described, using trapped ions or quantum dots. The Bristol Centre is patenting key aspects of the technology, and plans to offer it under licence; Nokia and Toshiba (who are manufacturing devices for the Bristol group) are already working on the development of photonic chips based on the Bristol breakthrough. The Bristol team are enthusiastic – perhaps understandably over-enthusiastic – about the possibilities. Speaking in 2012, Mark Thompson, a leading member of the Bristol group, said that single-purpose 'computers' for use in cryptography should be available (at least to customers with deep pockets) within three years, because they only need one pair of entangled photons, or two pairs allowing for the teleportation trick; that is plausible enough. Soon after, he expects to have chips each using twenty pairs of working photons to produce 10 qubit computers that will be able to carry out some kinds of calculation faster than conventional computers;[21] other researchers regard this as the limit of practical possibilities with existing photon sources. But Thompson also expects to have systems using hundreds of qubits that can be applied to specific tasks – such as determining the shape of a molecule in what is known as protein folding, a key step in the development of new drugs – 'within ten years'. If these projections are well founded, they make quantum photonics very much the front runner in the quantum computer race at the time of writing, late in 2012; but perhaps the more ambitious forecasts should be taken with a pinch of salt.

As with conventional chips, though, if a working proto-type can be developed, virtually all of the expense attaches to designing and manufacturing the first chip. Once you have set up the machinery to make one, you can make a million at marginal extra cost. This raises the real possibility, for example, of incorporating quantum technology into smart-phones, although maybe not on a ten-year timescale. Why bother? Well, although quantum computers can be used to crack conventional codes, they can be used to create codes that cannot be cracked even by conventional computers.[22] This doesn't just mean that celebrities and politicians could enjoy making phone calls without the fear of being hacked. It means that we could all administer our bank accounts or handle other sensitive data from our phones without worrying that the information might be captured by a third party and misused. Even if our computers and smartphones will, as I have explained, still have to use classical methods for dealing with most problems, the prospect of unhackable computers and smartphones is alone sufficient to justify the effort going into computing with quanta.

This brings my story to a pleasingly symmetrical con-clusion. It started with Alan Turing and the need to crack codes; it ends with Turing's heirs and the need for uncrackable codes. From Colossus to qubits, the story is essentially the same.

Coda

A Quantum of Discord

The story of quantum computing is usually told, as I have told it here, in terms of small numbers of quantum entities, entanglement, and superposition. Everything I have told you about this approach to computing with quanta is correct. But there may be something else going on, something which could involve a different approach to computing with quanta, and may also provide insight into the foundations of quantum mechanics. You may, perhaps, have noticed something odd about the very first quantum computer I described, which was indeed the first one to apply Shor's algorithm successfully – the NMR technique which used a thimbleful of liquid to factorize the number 15. Several physicists pointed out that in a liquid at room temperature it is not possible to maintain entanglement and superpositions; the nuclear spins are shaken up by interactions between particles and cannot be neatly aligned. And yet, the NMR systems work! Whatever the intention of the people who designed those experiments, they must, as the experimenters themselves came to acknowledge, be working for another reason. What was it that gave these systems the power to run Shor's algorithm?

The answer seems to be a phenomenon known as discord, a term introduced to quantum physics by Wojciech Zurek, of the Los Alamos National Laboratory, New Mexico, in 2000. Quantum discord provides a measure of quantum correlations, and in particular of how much a system is disturbed when it is measured and information is obtained from it. Everyday objects, including classical computers, are unchanged when they are observed, so they have zero discord; quantum systems, as we have seen, are affected by being forced to collapse into specific states, so they have positive values of discord. The person credited with applying this idea to quantum computing is Animesh Datta, of the University of Oxford. He built on work by Emanuel Knill, now at NIST, and Raymond Laflamme, now at the University of Waterloo, Canada. They had raised the question of what would happen if a qubit in a 'mixed' state (that is, not 0 or 1 but some messier state, say, one-third 0 and two-thirds 1) were sent through an entangling gate with a 'pure' qubit, in a definite state (0 or 1). Mixed qubits cannot be used for entangling; but they can interact with pure qubits, and it turned out that such a quantum interaction between the mixed and pure qubits, which is described mathematically by discord, can be used in computation. In other words, instead of preparing and controlling two pure qubits on their way through the network of gates that make up a quantum computer, only a single carefully controlled pure qubit is needed, while the other one can be knocked about by its interactions with the surrounding environment. This seemed to be the reason for the success of the early NMR experiments. As long as some of the nuclei were playing ball by being aligned in accordance with the expectations of the researchers, it didn't matter that most were being jostled out of the pure states that had seemed so important.

Experimenters seized on the idea of discord, and soon (in 2008) showed that it works on a small scale. In a classic example of the scientific method, one of the first tests was carried out by Andrew White, at the University of Queensland, who was convinced that it would not work, and expected to prove Datta wrong. But it did, so he had to change his mind about the value of discord in computation. In science, it isn't what you believe, or wish, to be true that matters; it's what the experiments tell you.

Discord has been likened to the hiss of background noise that you get from an AM radio tuned to a weak station. The bizarre thing is that putting noise into the computer system (White worked with polarized and mixed photons) makes it more powerful. Including information about the noise provides more computational power than if it is ignored. This is great news for experimenters, and engineers, since it is easier for them to work with noisy systems than to go to the trouble of maintaining everything in a pure state. But it is much harder for the theorists to analyse such systems.

Another analogy has been made with aircraft. Ever since the Wright brothers, we have known that it is possible for machines heavier than air to fly; but aerodynamicists are still unable to come up with a precise mathematical explanation of how such flying machines work. This does not, however, discourage them from using such machines! So while one line of research into computing with quanta progresses on the pure principles described in Chapter 6, we may be entering an era in which there is a rival engineering approach, the 'suck it and see' technique, which operates by developing machines that work, even if nobody quite knows how they work. It now seems that this is what was going on in the early NMR experiments. They do involve quantum interactions, but not the quantum interactions the theorists thought were involved. It was, you might say, 'quantum computing, Jim, but not as we know it'.

Notes

Introduction: Computing with Quantum Cats

1   The discussion here borrows from the one in my book *In Search of the Multiverse*, which I could not improve upon.

Chapter 1: Turing and the Machine

1   She was christened Ethel Sara, but preferred Sara.

2   Information about Alan's childhood, here and later, from Sara Turing's memoir, *Alan M. Turing*.

3   This proved more useful than earlier prizes he had received while at school, mostly in the form of literary works. These books are now in the museum at Bletchley Park, and show little sign of ever having been read.

4   Gian-Carlo Rota, quoted by Leavitt, *The Man Who Knew Too Much*.

5   Quoted by Dyson, *Turing's Cathedral*.

6   Letter quoted by Hodges, *Alan Turing*.

7   I oversimplify greatly; for the full story, see Jack Copeland and others, *Colossus*.

8   Through his contacts with GC&CS, Turing was already aware of, and had done a considerable amount of work on, Enigma while still in Cambridge, with occasional visits to GC&CS: see Copeland (ed.), *The Essential Turing*.

9   Simon Singh, *The Code Book*.

10   The process involved, apart from the encipherment, converting analogue signals into digital form and back again, as in the conversion of music into MP3 files and back again. My own Turing machine is doing this as I write.

11   Of course, messages could also be sent in real time, letter by letter as the operators typed; they would use this method to make contact before running a prepared paper tape through the transmitter.

12   Flowers, quoted in Copeland and others, *Colossus*.

13   From the initials of Women's Royal Naval Service.

14   Copeland and others, *Colossus*.

15   http://www.AlanTuring.net/tunny_report.

16   Not that Turing cared much for such honours; he kept the medal in a tin box along with an assortment of nails, screws and other odds and ends.

17   Quoted in Lavington (ed.), *Alan Turing and His Contemporaries*.

18   His essay 'My Brother Alan' appears in the centenary edition of Sara Turing's book.

19   Quoted in Lavington (ed.), *Alan Turing and His Contemporaries*.

20   You can hear them at http://www.digital60.org/media/mark_one_digital_music.

21   Turing's secretary in Manchester recalls that his handwriting was so bad that often he could not read it himself, and had to ask her to decipher it.

22   He wrote to his brother after the arrest to seek his help and advice. The first sentence of the letter read: 'I suppose you know I am a homosexual.' But 'I knew no such thing,' says John. Turing's colleague Donald Michie has commented (see Sara Turing, *Alan M. Turing*) that Alan was 'so child-like and fundamentally good as to make him a very vulnerable person in a world so largely populated by self-seekers'.

Chapter 2: Von Neumann and the Machines

1   This wasn't unusual. As late as the 1920s, the going rate for a knighthood in the UK was about £5,000.

2   Interview quoted by Dyson.

3   Eugene Wigner, who would later be awarded the Nobel Prize for his work in theoretical nuclear physics, was a year ahead of von Neumann at the same school.

4   Interview with George Dyson, *Darwin among the Machines*.

5   We don't know even today what he had been up to in England, but he later wrote (see Dyson, *Darwin among the Machines*) that 'I

received in that period a decisive impulse which determined my interest in computing machines'. Given that he already knew Turing, I think we can put two and two together.

6   See Hargittai, *Martians of Science*.

7   Quoted by Truesdell, *The Development of Punch Card Tabulations*.

8   Watson was the head of IBM at the time.

9   Bizarrely, Zuse's application for a patent on the Z3 was turned down on the basis that it 'lacked inventiveness'.

10   This is important in radar.

11   Interview in Abramson, *Zworykin, Pioneer of Television*.

12   Goldstine, *The Computer from Pascal to von Neumann*.

13   This method of sharing data and programs became common; two decades later, the data I used for the work on my PhD, carried out on an IBM 360 machine, was sent from California to Cambridge in the same form, although requiring rather fewer cards.

14   Quoted by Dyson, *Darwin among the Machines*. Goldstine, always more generous to his mentor, says that it was von Neumann 'who took the raw idea and perfected it', and who 'crystallized thinking in the field of computers as no other person ever did' (*The Computer from Pascal to von Neumann*). True, but he could have been more generous in sharing the credit.

15   Goldstine, *The Computer from Pascal to von Neumann*.

16   Goldstine tells us that von Neumann 'had a profound concern for automata. In particular, he always had a deep interest in Turing's work.' Similar mathematical ideas to Turing's had been published in the same year as 'On Computable Numbers' by Emil Post, of the City College of New York. However, 'There is no doubt that von Neumann was thoroughly aware of Turing's work but apparently not of Post's' (Goldstine, *The Computer from Pascal to von Neumann*).

17   Mostly because they are outside my area of competence.

18   Von Neumann's thoughts on self-reproducing automata were gathered in a posthumous book, *Theory of Self-Reproducing Automata*, edited by Arthur Burks, from which his words in this section are taken.

19   Originally a BBC TV series; now available in book form.

20   Quoted by Leavitt, *The Man Who Knew Too Much* (my emphasis).

21   McCarthy and Shannon (eds), *Automata Studies*.

First Interlude: Classical Limits

1   In 1997, Moore came up with another striking analogy. The total number of transistors on all of the chips manufactured by Intel in that year was roughly equal to the number of ants on Earth, about 100 million billion.
2   Actually, not so rapidly, as progress is beginning to slow down as the limits of Moore's Law are approached.
3   Available at http://www.zyvex.com/nanotech/feynman.html.

Chapter 3: Feynman and the Quantum

1   For the purposes of this book, the terms 'quantum mechanics' and 'quantum physics' are synonymous.
2   Mehra, *The Beat of a Different Drum*.
3   In the US system, 'senior' is essentially synonymous with third year undergraduate.
4   Mehra, *The Beat of a Different Drum*.
5   Actually, not quite at a definite point. Heisenberg also discovered that no quantum entity can exist at a definite point: there is always some uncertainty in its position. This is a real, intrinsic uncertainty, not the result of our imperfect measurements; the electron itself does not 'know' precisely where it is at any moment in time. But we can ignore this Heisenberg uncertainty for the present discussion, because it is very small in these circumstances.
6   Feynman, *Lectures on Physics*, vol. 3.
7   Feynman, *Lectures on Physics*, vol. 3.
8   Reprinted in the volume edited by Laurie Brown, *Selected Papers of Richard Feynman*.
9   Honesty compels me to say that this is not quite like the kick of a rifle, but the image is a helpful one.
10   Strictly, the equations now known as Maxwell's equations are a tidied-up version of what he discovered.
11   Dirac's paper is reprinted at the end of the book *Feynman's Thesis*; his insight can also be found in his textbook *The Principles of Quantum Mechanics*, published by Clarendon Press, Oxford in 1935 (2nd edn), but this is too technical for the non-specialist.
12   It is always important to stress that no such experiment has ever been carried out with a real cat.

13   See Leff and Rex, *Maxwell's Demon*; this also includes the 1961 paper.
14   See Leff and Rex, *Maxwell's Demon*, ch. 4.8; this collection also includes the 1973 paper.
15   See Wright, *Three Scientists and their Gods*.
16   Feynman, *Lectures on Computation*.
17   Reprinted in Feynman, *Lectures on Computation*.
18   The Controlled NOT, or CNOT, gate is particularly important in quantum computation, playing a similar role to Fredkin gates in classical computation. It is possible to build any logic circuit using combinations of CNOT gates. See Chapter 5.

Chapter 4: Bell and the Tangled Web

1   A term first used, in this sense, by Schrödinger in 1935, and presented in the same paper as his cat puzzle. He described the 'entangling' of two quantum wave functions as 'the characteristic of quantum mechanics [that] enforces its entire departure from classical lines of thought'.
2   Louis inherited the title 'Duc' when his brother died in 1960.
3   In a talk celebrating de Broglie's 90th birthday in 1982: CERN TH 3315. A version of this is available in Bell, *Speakable and Unspeakable in Quantum Mechanics*.
4   Bell, *Speakable and Unspeakable in Quantum Mechanics*.
5   Somewhat more surprisingly, it wasn't even included in the collection of papers about quantum theory and measurement edited by Wheeler and Zurek and published in 1983.
6   See Wheeler and Zurek, *Quantum Theory and Measurement*.
7   In a letter to Max Born.
8   1935 was, of course, also the year that Schrödinger published his cat 'paradox', discussed in Chapter 3. But at that time Einstein and Schrödinger were essentially the only ones questioning the Copenhagen Interpretation.
9   Much later, Einstein expressed this idea forcefully in a conversation with Abraham Pais. 'Do you really believe', he asked, 'that the Moon exists only when I look at it?' See *Reviews of Modern Physics*, Vol. 51 (1979), p. 863.
10   Because, of course, the effect is not local, but at a distance.
11   See Hiley and Peat (eds), *Quantum Interpretations*.

12 Bell quotes in this section from Bernstein, *Quantum Profiles*, unless otherwise indicated.
13 My emphasis.
14 And also in my own *Schrödinger's Kittens*.
15 See Davies and Brown, *The Ghost in the Atom*.
16 See Whitaker, *The New Quantum Age*.
17 See Niels Bohr archive, http://www.aip.org/history/ohilist/25643.html.
18 In the Bohr archive interview, in 2002, Shimony described Bell as 'awesome', saying: 'He was discontented with a solution that was hand waving. He was one of the most rigorously honest men ever.'
19 See Niels Bohr archive, http://www.aip.org/history/ohilist/25096.html.
20 See Bell, *Speakable and Unspeakable in Quantum Mechanics*, ch. 16.
21 See Bertlmann and Zeilinger (eds), *Quantum (Un)speakables*.
22 Davies and Brown, *The Ghost in the Atom*.

Second Interlude: Quantum Limits
1 That's one hundred billion billion Planck lengths in everyday language.
2 Feynman, *Lectures on Computation*.

Chapter 5: Deutsch and the Multiverse
1 Bell, *Speakable and Unspeakable in Quantum Mechanics*, ch. 15.
2 For sources and more detail on this and the Multiverse idea in general, see my book *In Search of the Multiverse*.
3 Bell, *Speakable and Unspeakable in Quantum Mechanics*, ch. 20.
4 Bell, *Speakable and Unspeakable in Quantum Mechanics*, ch. 20.
5 See Woolf (ed.), *Some Strangeness in the Proportion*.
6 Remember that FAPP it doesn't matter which version of quantum mechanics you choose to work with; see my book *Schrödinger's Kittens*. It is possible to 'explain' quantum computing in other interpretations, but MWI seems the most natural to me, so I will use it from here on.
7 In 1998 he was awarded the Dirac Prize of the Institute of Physics, in 2005 he received the Edge of Computation Science Prize worth $100,000, and in 2008 he was elected a

Fellow of the Royal Society, with no cheque but plenty of kudos.

8   See Davies and Brown, *The Ghost in the Machine*. Deutsch elaborated on the idea in his book *The Fabric of Reality*.

9   The Copenhagen Interpretation was still conventional in 1982.

10   If you want them, see *The Beginning of Infinity*.

11   'Quantum theory, the Church–Turing principle and the universal quantum computer', in *Proceedings of the Royal Society*, Vol. 400 (1985), pp. 97–117.

12   His emphasis.

13   There is no significance in the choice of decimal numbers; any different decimals will do.

14   In practice the coding would be more sophisticated than this, as with Enigma, but this simple example makes the point.

15   Using the best computers available at the time of writing, summer 2012.

16   Long ago, I used this application of the technique in the work for my PhD thesis.

17   His emphasis.

18   In fairness, although I personally like the MWI, I should acknowledge that proponents of alternative interpretations are equally convinced that only 'their' view is right. See my book *Schrödinger's Kittens*. As Winfried Hensinger commented to me, 'How can you quantify weirdness? We should not forget our intuition is based in a classical world, so it will always mislead us in any interpretation of quantum physics.'

19   It has even been suggested that our Universe is a simulation running on a quantum computer; see *In Search of the Multiverse*.

20   The P stands for 'polynomial time'.

21   For 'nondeterministic polynomial time'.

22   This is a version of the 'travelling salesman problem'.

23   That is, as still as is allowed by quantum uncertainty.

24   Not to be confused with the quantum pioneer Wolfgang Pauli.

25   In the sense of being made in one piece.

26   $16 = 2^4$.

27   This really has been done. See next chapter. It has also been done with ions, on a much smaller scale.

28   I should mention that Steane is not a proponent of the MWI and that this remark is my own.

29   The spin is a property of the nucleus, but sometimes people – myself included – refer sloppily to the spin of the atom.

Chapter 6: Turing's Heirs and the Quantum Machines

1   See his contribution to Bernstein and Lo (eds), *Scaleable Quantum Computers*. Also arXiv:quant-ph/0002077v3, 'The Physical Implementation of Quantum Computation'.
2   All the other techniques require some kind of breakthrough in terms of the physics to become viable, but every aspect of the ion trap technique has been tried and tested. The problems that remain are engineering problems, and funding. Given money and time, we can be sure an ion-trap quantum computer will be built. The question is whether a physics breakthrough will enable another technique to get there first.
3   *The Guardian*, 30 September 2001.
4   Technically, superconducting Cooper pairs of electrons.
5   Essentially, phase tells you if waves are in step with one another.
6   *New York Times*, 28 February 2012.
7   See more at: http://www.theage.com.au/technology/sci-tech/tiny-dot-speeds-hitech-future-20100524-w4bi.html#ixzz27fftbDQE.
8   It's convenient to think of a single electron occupying a quantum dot; in practice, it may be that there are several electrons in each quantum dot, with one more in one dot than the other.
9   At least, for electrons and other so-called 'spin half' systems. I shall return to this shortly.
10   *Nature*, Vol. 489 (27 September 2012), pp. 541–5.
11   *Science*, Vol. 336, No. 6086 (8 June 2012).
12   Specifically, phosphorus-39.
13   Protons and neutrons themselves are composed of quarks, but happily we do not have to go down to that level of detail here.
14   Fluids are preferred to solids because they have no special structure to complicate calculations, unlike crystals, for example.
15   The tortoise put on a bit of a spurt itself in 2009, when a team at NIST developed a two-qubit system that could carry out any of 160 different operations on demand, making it in effect the world's first programmable quantum computer.
16   *Scientific American*, August 2008.

17   Europe, Canada and Japan have similar plans in the pipeline.

18   Richard Feynman once said that he would like to write a popular book, but couldn't decide whether it should be entitled *Fun with Fysics* or *Phun with Physics*. I have (barely) resisted the equivalent temptation here.

19   Never forget, though, that those 'standard industrial techniques' require the use of computers to control the operations.

20   *Science Express* (online), 27 March 2008.

21   This is (slightly) optimistic; most people I spoke to who work in the field think you would need around 50 qubits to solve problems a classical computer cannot do.

22   See my book *Schrödinger's Kittens*.

Sources and Further Reading

Books

Abramson, Albert, *Zworykin, Pioneer of Television* (Chicago: University of Illinois Press, 1995)

Aspray, William, *John von Neumann and the Origins of Modern Computing* (Cambridge, Mass.: MIT Press, 1991)

Bell, John, *Speakable and Unspeakable in Quantum Mechanics* (Cambridge: Cambridge University Press, 1987)

Bergin, Thomas (ed.), *50 Years of Army Computing* (Aberdeen, Md., Army Research Laboratory, 1999)

Bernstein, Jeremy, *Quantum Profiles* (Princeton: Princeton University Press, 1991)

Bernstein, S. L. and H.-K. Lo (eds), *Scaleable Quantum Computers* (Berlin: Wiley-VCH, 2001)

Bertlmann, Reinhold and Anton Zeilinger (eds), *Quantum (Un)speakables* (Berlin: Springer, 2002)

Bohm, David, *Quantum Theory* (Englewood Cliffs, NJ: Prentice-Hall, 1951; Dover edition, 1989)

Born, Max, *Natural Philosophy of Cause and Chance* (Oxford: Clarendon Press, 1949)

Born, Max, *The Born–Einstein Letters* (London: Macmillan, 1971)

Bronowski, Jacob, *The Ascent of Man* (London: BBC Books, 2011; first pub. 1973)

Brown, Julian, *Minds, Machines, and the Multiverse* (New York: Simon & Schuster, 2000)

Brown, Laurie (ed.), *Selected Papers of Richard Feynman* (Singapore: World Scientific, 2000)

Calvocoressi, Peter, *Top Secret Ultra*, rev. edn (London: Baldwin, 2011)

Copeland, Jack (ed.), *The Essential Turing* (Oxford: Oxford University Press, 2004)

Copeland, Jack, and others, *Colossus: The Secrets of Bletchley Park's Codebreaking Computers* (Oxford: Oxford University Press, 2006)

Davies, Paul and Julian Brown, *The Ghost in the Atom* (Cambridge: Cambridge University Press, 1986)

Deutsch, David, *The Fabric of Reality* (London: Allen Lane, 1997)

Deutsch, David, *The Beginning of Infinity* (London: Allen Lane, 2011)

DeWitt, Bryce and Neil Graham (eds), *The Many-Worlds Interpretation of Quantum Mechanics* (Princeton, NJ: Princeton University Press, 1973)

Dodgson, Charles, *The Game of Logic* (London: Macmillan, 1887)

Dodgson, Charles, *Symbolic Logic*, 3rd edn (London: Macmillan, 1896)

Dyson, George, *Darwin among the Machines* (London: Allen Lane, 1998)

Dyson, George, *Turing's Cathedral* (London: Allen Lane, 2012)

Eckert, Wallace, *Punched Card Methods in Scientific Computation* (New York: Thomas J. Watson Astronomical Computing Bureau, Columbia University, 1940)

Farmelo, Graham, *The Strangest Man* (London: Faber & Faber, 2009)

Feynman, Richard, *Lectures on Physics*, vol. 3 (Reading, Mass.: Addison-Wesley, 1965)

Feynman, Richard, *Lectures on Computation*, ed. Anthony Hey and Robin Allen (London: Penguin, 1999)

Feynman, Richard, *Feynman's Thesis*, ed. Laurie Brown (Singapore: World Scientific, 2005)

Feynman, Richard and Ralph Leighton, *Surely You're Joking, Mr. Feynman?* (New York: Norton, 1985)

Gilder, Louisa, *The Age of Entanglement* (New York: Knopf, 2008)

Goldstine, Herman, *The Computer from Pascal to von Neumann* (Princeton, NJ: Princeton University Press, 1972)

Gribbin, John, *Schrödinger's Kittens and the Search for Reality* (London: Weidenfeld & Nicolson, 1995; pb Phoenix, 1996)

Gribbin, John, *In Search of the Multiverse* (London: Allen Lane, 2009)

Gribbin, John, *Erwin Schrödinger and the Quantum Revolution* (London: Bantam, 2012)

Gribbin, John and Mary Gribbin, *Richard Feynman: A Life in Science* (London: Viking, 1997)

Hargittai, István, *Martians of Science* (Oxford: Oxford University Press, 2006)

Heisenberg, Werner, *Physics and Beyond* (London: Allen & Unwin, 1971)

Herbert, Nick, *Quantum Reality* (New York: Anchor Press/Doubleday, 1985)

Hey, Anthony (ed.), *Feynman and Computation* (Reading, Mass.: Perseus, 1999)

Hiley, Basil and David Peat (eds), *Quantum Interpretations* (London: Routledge & Kegan Paul, 1987)

Hodges, Andrew, *Alan Turing* (London: Burnett/Hutchinson, 1983)

Hoyle, Fred and John Elliott, *A for Andromeda* (London: Souvenir, 2001)

Lavington, Simon (ed.), *Alan Turing and His Contemporaries* (Swindon: BCS, 2012)

Leff, Harvey and Andrew Rex, *Maxwell's Demon: Entropy, Information and Computing* (Bristol: Adam Hilger, 1990)

Leff, Harvey and Andrew Rex, *Maxwell's Demon 2: Entropy, Classical and Quantum Information, Computing* (Bristol: Institute of Physics, 2003)

Leavitt, David, *The Man Who Knew Too Much* (London: Weidenfeld & Nicolson, 2006)

McCarthy, John and Claude Shannon (eds), *Automata Studies* (Princeton, NJ: Princeton University Press, 1956)

McKay, Sinclair, *The Secret Life of Bletchley Park* (London: Aurum, 2010)

Macrae, Norman, *John von Neumann* (New York: Random House, 1993)

Mehra, Jagdish, *The Beat of a Different Drum* (Oxford: Clarendon Press, 1994)

Mermin, David, *Boojums All the Way Through* (Cambridge: Cambridge University Press, 1990)

Mermin, David, *Quantum Computer Science* (Cambridge: Cambridge University Press, 2007)

Neumann, John von, *The Mathematical Foundations of Quantum Mechanics*, trans. R. T. Beyer (Princeton, NJ: Princeton University Press, 1955; first pub. in German, 1932)

Neumann, John von, *The Computer and the Brain* (New Haven, Conn.: Yale University Press, 1958)

Neumann, John von, *Theory of Self-Reproducing Automata*, ed. Arthur Burks (Urbana: University of Illinois Press, 1966)

Penrose, Oliver, *Foundations of Statistical Mechanics* (Oxford: Pergamon, 1970)

Petzold, Charles, *The Annotated Turing* (Indianapolis: Wiley, 2008)

Rieffel, Eleanor and Wolfgang Polack, *Quantum Computing* (Cambridge, Mass.: MIT Press, 2011)

Singh, Simon, *The Code Book* (London: Fourth Estate, 1999)

Truesdell, Leon, *The Development of Punch Card Tabulations in the Bureau of the Census, 1890–1940* (Washington DC: US Government Printing Office, 1965)

Turing, Sara, *Alan M. Turing*, centenary edn (Cambridge: Cambridge University Press, 2012)

Wheeler, John and Wojciech Zurek (eds), *Quantum Theory and Measurement* (Princeton, NJ: Princeton University Press, 1983)

Whitaker, Andrew, *The New Quantum Age* (Oxford: Oxford University Press, 2012)

Willans, Geoffrey and Ronald Searle, *Molesworth*, new edn (London: Penguin Modern Classics, 2000)

Woolf, Harry (ed.), *Some Strangeness in the Proportion: Centennial Symposium to Celebrate the Achievements of Albert Einstein* (Reading, Mass.: Addison-Wesley, 1981)

Wright, Robert, *Three Scientists and their Gods* (New York: Times Books, 1988)

Zuse, Konrad, *The Computer: My Life* (Berlin: Springer, 1993; first pub. in German as *Der Computer*, 1970)

Web

http://libweb.princeton.edu/libraries/firestone/rbsc/finding_aids/mathoral/pmcxrota.htm

http://www.sussex.ac.uk/physics/iqt/virtualtour.html

Picture Acknowledgements

Every effort has been made to trace the copyright holders of photos reproduced in the book. Copyright holders not credited are invited to get in touch with the publishers.

Photos in the text
7: The Granger Collection/Topfoto; 9: King's College Library, Cambridge/AMT/K/7/12; 53: Time & Life Pictures/Getty Images; 97: provided with kind permission of Dr Tonomura; 99: CERN/Science Photo Library; 135: © 1982 CERN; 181: Winfried Hensinger, University of Sussex; 183: Lulie Tanett; 226: Getty Images

Illustrations section
Credits are listed clockwise on each spread, starting from the top left.
Letter from Alan Turing, 1 April 1923: King's College Library, Cambridge/AMT/K/1/1, © P. N. Furank; Alan Turing on Waterloo Station, c. 1926 (detail): King's College Library, Cambridge/AMT/K/7/3; Alan Turing finishing a race: National Physical Laboratory © Crown copyright/ Science Photo Library; Colossus, 1943 and code-breakers, Bletchley Park, c. 1942: SSPL via Getty Images; Hut 3, Bletchley Park: © Edifice/Corbis

ENIAC: Associated Press; John Hoyle: BBC Photo Library; bio-wall: Philippe Plailly/ Science Photo Library; advertisement for the Bendix G-15 computer: courtesy of the Computer History Museum; J. Robert Oppenheimer and John von Neumann: Emilio Segre Visual Archives/American Institute of Physics/Science Photo Library

Fifth Solvay Physics Conference, Brussels, 1927: SSPL via Getty Images; John Stewart Bell, 1989: Corbin O'Grady Studio/Science Photo Library; David Bohm, 1971: Getty Images; Alain Aspect: Österreichische Zentralbibliothek für Physik (Austrian Central Library for Physics); double-slit refraction: GIPhotostock/Science Photo Library

Hans Georg Dehmelt: Emilio Segre Visual Archives/ American Institute of Physics/Science Photo Library; Brian Josephson, 23 November 1973: PA/PA Archive/Press Association Images; Gary Kasparov vs a computer, 3 May 1997: AFP/Getty Images; quantum cryptography equipment: Volker Steger/Science Photo Library; MRI scanner, 2010: Boston Globe via Getty Images; Nobel laureates David J. Wineland and Serge Haroche, December, 2012: AFP/Getty Images; David Wineland adjusting an ultraviolet laser, 2003: © Geoffrey Wheeler/National Institute of Standards and Technology

Ion trap laboratory and chip: Winfried Hensinger, University of Sussex

Index

INDEX